Robust Microelectronic Devices

Robust Microelectronic Devices

Editor

Michael Waltl

MDPI • Basel • Beijing • Wuhan • Barcelona • Belgrade • Manchester • Tokyo • Cluj • Tianjin

Editor
Michael Waltl
Institute for Microelectronics
Austria

Editorial Office
MDPI
St. Alban-Anlage 66
4052 Basel, Switzerland

This is a reprint of articles from the Special Issue published online in the open access journal *Crystals* (ISSN 2073-4352) (available at: https://www.mdpi.com/journal/crystals/special_issues/robust_devices).

For citation purposes, cite each article independently as indicated on the article page online and as indicated below:

LastName, A.A.; LastName, B.B.; LastName, C.C. Article Title. *Journal Name* **Year**, *Volume Number*, Page Range.

ISBN 978-3-0365-3337-7 (Hbk)
ISBN 978-3-0365-3338-4 (PDF)

© 2022 by the authors. Articles in this book are Open Access and distributed under the Creative Commons Attribution (CC BY) license, which allows users to download, copy and build upon published articles, as long as the author and publisher are properly credited, which ensures maximum dissemination and a wider impact of our publications.
The book as a whole is distributed by MDPI under the terms and conditions of the Creative Commons license CC BY-NC-ND.

Contents

About the Editor vii

Michael Waltl
Editorial for the Special Issue on Robust Microelectronic Devices
Reprinted from: *Crystals* **2022**, *12*, 16, doi:10.3390/cryst12010016 1

Yoanlys Hernandez, Bernard Stampfer, Tibor Grasser, and Michael Waltl
Impact of Bias Temperature Instabilities on the Performance of Logic Inverter Circuits Using Different SiC Transistor Technologies
Reprinted from: *Crystals* **2021**, *11*, 1150, doi:10.3390/ cryst11091150 5

Dingwang Yu, Yanfei Dong, Youde Ruan, Guochao Li, Gaosheng Li, Haomin Ma, Song Deng and Zhenpeng Liu
Photo-Excited Switchable Terahertz Metamaterial Polarization Converter/Absorber
Reprinted from: *Crystals* **2021**, *11*, 1116, doi:10.3390/cryst11091116 17

Paulina Valencia-Gálvez, Daniela Delgado, María Luisa López, Inmaculada Álvarez-Serrano, Silvana Moris and Antonio Galdámez
AgSn[$Bi_{1-x}Sb_x$]Se_3: Synthesis, Structural Characterization, and Electrical Behavior
Reprinted from: *Crystals* **2021**, *11*, 864, doi:10.3390/cryst11080864 27

Pei-Te Lin, Jia-Wei Chang, Syuan-Ruei Chang, Zhong-Kai Li, Wei-Zhi Chen, Jui-Hsuan Huang, Yu-Zhen Ji, Wen-Jeng Hsueh and Chun-Ying Huang
A Stable and Efficient Pt/n-Type Ge Schottky Contact That Uses Low-Cost Carbon Paste Interlayers
Reprinted from: *Crystals* **2021**, *11*, 259, doi:10.3390/cryst11030259 39

Maximilian W. Feil, Andreas Huerner, Katja Puschkarsky, Christian Schleich, Thomas Aichinger, Wolfgang Gustin, Hans Reisinger, and Tibor Grasser
The Impact of Interfacial Charge Trapping on the Reproducibility of Measurements of Silicon Carbide MOSFET Device Parameters
Reprinted from: *Crystals* **2020**, *10*, 1143, doi:10.3390/cryst10121143 49

Saskia Schimmel, Daisuke Tomida, Tohru Ishiguro, Yoshio Honda, Shigefusa Chichibu and Hiroshi Amano
Numerical Simulation of Ammonothermal Crystal Growth of GaN—Current State, Challenges, and Prospects
Reprinted from: *Crystals* **2021**, *11*, 356, doi:10.3390/cryst11040356 63

Milan Ťapajna
Current Understanding of Bias-Temperature Instabilities in GaN MIS Transistors for Power Switching Applications
Reprinted from: *Crystals* **2020**, *10*, 1153, doi:10.3390/cryst10121153 93

About the Editor

Michael Waltl's overall scientific focus is on the robustness of microelectronic devices and circuits. In this field, he investigates reliability issues—characterization and modeling—in semiconductor devices and circuits. His research covers evaluating bias temperature instabilities in silicon devices, devices employing wideband-gap materials, and transistors built on novel 2D materials. Dr. Waltl also studies hot carrier degradation and stress-induced leakage currents of various transistor technologies. Another research pillar of Dr. Waltl is the high performant operation of integrated circuits built from scaled devices, where he investigates the impact of defects on the performance of circuits. Additionally, radiation hardening of various technologies, i.e. SRAM cells, and integrated electronic circuits, lies in his research focus. Furthermore, Dr. Waltl has a strong background in measurement technology and is leading the development of novel characterization tools and techniques. Dr. Waltl is the co-author or author of over 100 articles in journals and conference proceedings (h-index 23). He is the director of the Christian Doppler Laboratory for single defect spectroscopy in semiconductor devices and leads the device characterization laboratory at the IuE at the TUW. He is the (co-)recipient of various best paper awards (IIRW2014, DRC2019, IIRW2019, IEDM2019, etc.) and serves on the technical program and management committee of international conferences and workshops. Based on his expertise, Dr. Waltl is regularly invited as a reviewer of numerous renowned Journals, including *IEEE TED, Microelectronics Reliability, Journal of Applied Physics*, and many more.

Editorial

Editorial for the Special Issue on Robust Microelectronic Devices

Michael Waltl

Christian Doppler Laboratory for Single-Defect Spectroscopy, Institute for Microelectronics, TU Wien, Gusshausstrasse 27-29, 1040 Vienna, Austria; waltl@iue.tuwien.ac.at

Citation: Waltl M. Editorial for the Special Issue on Robust Microelectronic Devices. *Crystals* 2022, 12, 16. https://doi.org/10.3390/cryst12010016

Received: 17 December 2021
Accepted: 19 December 2021
Published: 23 December 2021

Publisher's Note: MDPI stays neutral with regard to jurisdictional claims in published maps and institutional affiliations.

Copyright: © 2021 by the author. Licensee MDPI, Basel, Switzerland. This article is an open access article distributed under the terms and conditions of the Creative Commons Attribution (CC BY) license (https:// creativecommons.org/licenses/by/ 4.0/).

Integrated electronic circuits have influenced our society in recent decades and become an indispensable part of our daily lives. To maintain this development and to ensure the benefits therefrom for decades to come, continuous further development of electronic chips is necessary. These developments include improving their performance and universality as well as exploiting the full potential of microelectronic technologies. It is also noteworthy that with the increasing number of applications for microelectronic systems, the associated demands on the individual components are also growing. A few challenges facing semiconductor industry research are the continuous optimization of integrated circuits towards higher operating frequencies and increasing the number of devices per chip area, while reducing power consumption. In addition to the abovementioned improvements, integrated electronic chips must also function robustly and reliably under various operating conditions, such as at low (cryogenic) or high temperatures, humidity, and many more.

The robustness, i.e., the high performance and reliable function, of microelectronic devices is the key for long-term failure-safe and stable operation of complex electrical circuits and applications. In recent decades, the performance and geometry of integrated devices have been continuously updated to improve performance, leading to new challenges that must be addressed. For instance, MOS transistors suffer from imperfections at the atomic level, which can emerge as electrically active sites—so-called defects. On the one hand, these defects are unavoidably introduced during device fabrication, and on the other hand, new defects can form during device operation at nominal bias conditions. The impact of these defects on the device performance itself manifests as a drift in the performance of the MOS transistors over time. In this context, the so-called bias temperature instability (BTI), which emerges as a drift of the threshold voltage of a transistor, is an essential criterion for determining the reliability of devices. Although significant emphasis has been placed on understanding this phenomenon and developing suitable models to explain the observed device performance degradation, the detailed physical mechanisms behind BTI are still controversial and debated.

In principle, BTI can be analyzed in two fundamentally different ways. The first is the investigation of large-area devices where continuous drifts of the threshold voltage, resulting from the superposition of the contributions of many defects, can be studied. This enables the calibration of analytical and compact models, essential for efficient circuit simulation. The second approach is to analyze the random telegraph noise (RTN) and investigate single defects by employing nanoscale devices. In doing so, the charge-trapping kinetics of single defects can be studied, which is vital for the development of physical defect models that further enable an accurate lifetime estimation under various operating conditions.

A continuous improvement of our understanding of BTI is not only essential for further optimization of silicon transistors, but also for the improvement of the performance of emerging technologies such as devices based on wide bandgap materials such as SiC or GaN, as well as for novel 2D transistors employing graphene, MoS_2, and many other 2D materials. In addition to the evolving challenges for the physical understanding of these observations, the study of novel material systems also poses a major challenge for suitable characterization techniques and measurement instruments, e.g., the requirement of high-speed measurement techniques (fast IDVG or fast CV methods), but also the need for ultra-low-noise systems, which allow us to investigate trap-assisted tunneling.

The publications published in this Special Issue are just as diverse as the field of research itself. In [1], the authors study the impact of BTI on the performance of logic inverter circuits using different SiC transistor technologies. As has been mentioned before, BTI is a severe reliability issue for all kinds of material systems, and thus, also for power electronics fabricated on SiC substrates. The authors report that defects in the structure of the SiC transistor can lead to a considerable change in signal propagation delay, which might become an issue at high operating frequencies. In [2], a photoexcited switchable THz metamaterial (MTM) polarization converter/absorber based on the incorporation of photoconductive silicon is designed and demonstrated. A key achievement with the discussed design is opening up a new field towards active switches and polarization manipulation with high performance in the THz regime. Potential applications of this novel structure compromise biological imaging, THz scanning, sensors, and more. The subsequent work in [3] studies the synthesis, structural characterization, and electrical behavior of AgSn[Bi$_{1-x}$Sb$_x$]Se$_3$, a candidate for thermoelectric materials. In [4], stable and efficient Ge Schottky contacts, based on low-cost carbon paste interlayers, are investigated. Schottky contacts are of particular importance for high-speed devices. Another important class of devices is discussed in [5]—i.e., SiC MOSFETs—and the impact of interfacial layer charge trapping on the device stability. In turn, the stability of a certain technology is vital for the reproducibility of measurements, and the authors report that experimental results strongly depend on the measurement scheme and the precise timing on a microsecond scale. Finally, two review papers have also been submitted to the special issue. In [6], the limitations and advances of simulating ammonothermal growth of bulk GaN is discussed. GaN is an important material system exhibiting a wide bandgap, making it attractive for powered electronic devices. The second review paper [7] outlines the robustness of GaN metal–insulator–semiconductor (MIS) transistors when used in power switching applications. It is noted that both negative BTI (NBTI) and positive (PBTI) are serious reliability concerns for this technology and require careful consideration when designing circuits and applications.

The Guest Editor would like to take the opportunity to thank all the researchers who contributed to this Special Issue by sharing their latest results and achievements. Furthermore, a special thank you goes to the reviewers for their time spent providing constructive feedback to the authors and helping to improve the quality of the published papers considerably. Finally, the Guest Editor wants to point out that there are a wide variety of further robustness issues of modern microelectronics devices and materials that are currently not fully understood. Therefore, it is essential to carefully look into these aspects if we are to ensure further improvements of integrated electronic applications for many years to come.

Funding: The financial support by the Austrian Federal Ministry for Digital and Economic Affairs, the National Foundation for Research, Technology and Development and the Christian Doppler Research Association is gratefully acknowledged.

Conflicts of Interest: The author declares no conflict of interest.

References

1. Hernandez, Y.; Stampfer, B.; Grasser, T.; Waltl, M. Impact of Bias Temperature Instabilities on the Performance of Logic Inverter Circuits Using Different SiC Transistor Technologies. *Crystals* **2021**, *11*, 1150, https://doi.org/10.3390/cryst11091150.
2. Yu, D.; Dong, Y.; Ruan, Y.; Li, G.; Li, G.; Ma, H.; Deng, S.; Liu, Z. Photo-Excited Switchable Terahertz Metamaterial Polarization Converter/Absorber. *Crystals* **2021**, *11*, 1116, https://doi.org/10.3390/cryst11091116.
3. Valencia-Gálvez, P.; Delgado, D.; López, M.L.; Álvarez Serrano, I.; Moris, S.; Galdámez, A. AgSn[Bi$_{1-x}$Sb$_x$]Se$_3$: Synthesis, Structural Characterization, and Electrical Behavior. *Crystals* **2021**, *11*, 864, https://doi.org/10.3390/cryst11080864.
4. Lin, P.T.; Chang, J.W.; Chang, S.R.; Li, Z.K.; Chen, W.Z.; Huang, J.H.; Ji, Y.Z.; Hsueh, W.J.; Huang, C.Y. A Stable and Efficient Pt/n-Type Ge Schottky Contact That Uses Low-Cost Carbon Paste Interlayers. *Crystals* **2021**, *11*, 259, https://doi.org/10.3390/cryst11030259.

5. Feil, M.W.; Huerner, A.; Puschkarsky, K.; Schleich, C.; Aichinger, T.; Gustin, W.; Reisinger, H.; Grasser, T. The Impact of Interfacial Charge Trapping on the Reproducibility of Measurements of Silicon Carbide MOSFET Device Parameters. *Crystals* **2020**, *10*, 1143, https://doi.org/10.3390/cryst10121143.
6. Schimmel, S.; Tomida, D.; Ishiguro, T.; Honda, Y.; Chichibu, S.; Amano, H. Numerical Simulation of Ammonothermal Crystal Growth of GaN—Current State, Challenges, and Prospects. *Crystals* **2021**, *11*, 356, https://doi.org/10.3390/cryst11040356.
7. Ťapajna, M. Current Understanding of Bias-Temperature Instabilities in GaN MIS Transistors for Power Switching Applications. *Crystals* **2020**, *10*, 1153, https://doi.org/10.3390/cryst10121153.

Article

Impact of Bias Temperature Instabilities on the Performance of Logic Inverter Circuits Using Different SiC Transistor Technologies

Yoanlys Hernandez [1], Bernhard Stampfer [2], Tibor Grasser [1] and Michael Waltl [2,*]

[1] Institute for Microelectronics, TU Wien, Gusshausstrasse 27-29, 1040 Vienna, Austria; hernandez@iue.tuwien.ac.at (Y.H.); grasser@iue.tuwien.ac.at (T.G.)
[2] Christian Doppler Laboratory for Single-Defect Spectroscopy, Institute for Microelectronics, TU Wien, Gusshausstrasse 27-29, 1040 Vienna, Austria; stampfer@iue.tuwien.ac.at
* Correspondence: waltl@iue.tuwien.ac.at

Abstract: All electronic devices, in this case, SiC MOS transistors, are exposed to aging mechanisms and variability issues, that can affect the performance and stable operation of circuits. To describe the behavior of the devices for circuit simulations, physical models which capture the degradation of the devices are required. Typically compact models based on closed-form mathematical expressions are often used for circuit analysis, however, such models are typically not very accurate. In this work, we make use of physical reliability models and apply them for aging simulations of pseudo-CMOS logic inverter circuits. The model employed is available via our reliability simulator Comphy and is calibrated to evaluate the impact of bias temperature instability (BTI) degradation phenomena on the inverter circuit's performance made from commercial SiC power MOSFETs. Using Spice simulations, we extract the propagation delay time of inverter circuits, taking into account the threshold voltage drift of the transistors with stress time under DC and AC operating conditions. To achieve the highest level of accuracy for our evaluation we also consider the recovery of the devices during low bias phases of AC signals, which is often neglected in existing approaches. Based on the propagation delay time distribution, the importance of a suitable physical defect model to precisely analyze the circuit operation is discussed in this work too.

Keywords: circuit reliability; pseudo-CMOS inverter circuits; SiC power MOSFETs; bias temperature instabilities; defect modeling; spice simulation

1. Introduction

Due to its outstanding properties, silicon carbide (SiC) is an excellent candidate for replacing conventional silicon-based power devices, especially for applications operating in harsh environments [1]. SiC MOSFETs offer a superior dynamic and thermal performance compared to traditional Silicon (Si) power MOSFETs. One key advantage of SiC as substrate material is that the achievable electric fields are around ten times higher than for their Si counterparts, which allows the design of MOSFETs with smaller on-resistance and smaller parasitic capacitance.

The features mentioned above for SiC MOSFETs have a positive impact, for instance, on the power dissipation for either lower or higher power levels [2] and play an essential role in the field of robust microelectronic devices. However, the performance of SiC MOSFETs is still below the theoretical limit of SiC, mainly due to the trapping of electrons in the channel region. Even though post-oxidation annealing has enabled the fabrication of high-quality SiC power transistors, the devices still suffer from a considerable number of imperfections at the semiconductor/insulator interface. The trapping of charge carriers at such defect states gives rise to a notable drift of the threshold voltage of SiC transistors, which is known as the bias temperature instability (BTI) [3,4]. As a consequence of BTI, an increase of V_{th} during the operation of the device introduces additional delays in circuits.

Furthermore, this may also lead to a higher on-resistance which negatively affects the power conversion efficiency of selected circuits [3]. Thus, the reliability of circuits employing SiC devices needs to be studied very carefully, especially for applications where the stable device performance and the lifetime are critical [5]. A particular challenge for high-power devices is that the drift of the threshold voltage in return can lead to an increase of the losses in the transistor and cause and increase in the operating temperature of the system [6]. Thus, threshold voltage stability is an essential issue for power devices and applications. To optimize the performance and analyze the circuit behavior under certain operating conditions, circuit simulations employing Spice simulators are a beneficial tool. In Spice simulators accurate compact models are the key components to precisely reproduce the electrical behavior of either devices or circuits. In our work we employ both physical device simulations and circuit simulations to evaluate the behavior of logic inverter circuits using different device technologies.

In more detail, we make use of 2nd and 3rd generation of commercially available SiC power MOSFETs provided by Cree, specifically the devices are C2M0280120D and C3M0065090J, respectively. All device kinds are fabricated on SiC substrate, however, their behavior it terms of drift of the threshold voltage is different. This is due to the fact that each device has been fabricated under different processing conditions which leads to a different trap distribution. As a consequence, our reliability model has to be calibrated to each device variant, i.e., to each device technology, individually. In the following we will refer to C3M0065090J as T1 and C2M0280120D as T2 to better comprehend the results. We also used the SCT10N120 SiC power MOSFET from STMicroelectronics. Likewise, we will identify this device from now on as T3. The device vendors provide the respective spice models of the transistors [7,8], which we employ in our simulations. The models have been calibrated using static and dynamic measurements. However, these models typically do not account for aging mechanisms such as BTI. To close this gap, we evaluate the impact of BTI on the device behavior under operation, e.g., the drift of the threshold voltage over time. Furthermore, we combine the provided models with our calibrated reliability simulations to thoroughly analyze the degradation of the performance of the inverter circuits over time.

2. Charge Trapping and Model Calibration

One of the most prominent stability issues of devices is the so -called BTI. This aging mechanism has been extensively studied in the literature for Si/SiO$_2$ [9–12], Si/HK [13–16] and SiC/SiO$_2$ material systems [17]. BTI typically evolves as a drift of the threshold voltage with operation time and is characterized at higher gate voltages to accelerate ΔV_{th} degradation. The origin of the observed ΔV_{th} lies in charge trapping at interface traps and oxide defects [18,19].

Using measure-stress-measure (MSM) experiments [18], one can study this phenomenon as well as the creation of new defects. Typically power-law functions have been widely used to reproduce the threshold voltage drift ΔV_{th} with stress time. However, using power-law-like formulas can lead to an erroneous prediction of BTI's impact on the device and circuit parameters. This simple formula does not account for the saturation of ΔV_{th} at high-stress times [19]. To ensure full accuracy in simulations, physical charge trapping models, for instance, the two-state defect model represented in Figure 1, should be preferably used [20,21]. This model is implemented in our open-source reliability simulator Comphy [21]. Our BTI simulator Comphy employs a two-state defect model based on the non-radiative multiphonon (NMP) theory. Moreover, it is used to calculate the charge capture and emission events at oxide defects to explain the ΔV_{th}. The most important parameters, as well as the expressions to determine the charge transitions times τ_{c} and τ_{e}, are shown in the configuration coordinate diagram in Figure 1. The principle idea of the model is that a charge carrier has to overcome the energy barrier depicted as E_{12} and E_{21} to change the charge state of a certain defect. In the semi-classical picture the barriers can be computed considering the intersection of the parabolas. Their relative position of the parabolas thereby depends on the applied gate bias and the energetic trap level of the defect. E_2 refers to the energy level of the carrier reservoir (e.g., conduction band edge or valence band edge), and E_1 describes the trap level of the defect. Other model parameters are the curvature of the parabolas c, the electron concentration n, the thermal velocity $v_{th,n}$, the capture cross-section σ, and the tunneling coefficient ν.

Using Comphy the experimental data from various transistor technologies can be accurately reproduced. In [17,19] we make use of this model to precisely explain charge trapping at DC and AC operating conditions, as shown in Figure 2, for the technologies investigated in this work. It has to be noted that especially the saturation of the threshold voltage drift behavior can be nicely explained by this model [19].

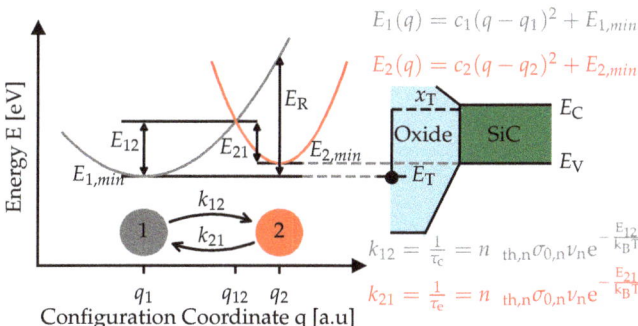

Figure 1. Schematic of the potential energy surfaces of our physical defect model and the corresponding model parameters required to calculate the charge trapping at defects. The model is based on non-radiative multi-phonon (NMP) theory to calculate the transition times at single defects corresponding to charge transfer reactions.

The calibration of our reliability simulation tool is based on two types of experiments [19,22]: (i) DC-MSM sequences for different bias applied at room temperature, see Figure 3 (**left**), and (ii) AC-MSM measurements to characterize the operation of the transistors in switching applications, see Figure 3 (**right**). In the last case, a short AC stress during a time around 100 ms is applied repeatedly. At the last AC cycle of the signal, the stress is interrupted ($t_{AC,interrupt}$). Then a ΔV_{th} recovery trace is extracted for a time equal to 10 ms considering a fixed I_D = 1 mA through the channel of the device.

To ensure full accuracy of our simulation framework the tools are calibrated considering both DC and short-term AC experimental data to derive the surface potential from a given gate voltage, doping concentration in the channel, the oxide thickness and work function difference for each technology. As can be seen in Figures 4–6, using our theoretical model can nicely replicate the device behavior for static and quasi-static operation conditions. This enables us now to look more detailed into the impact of BTI on different circuits.

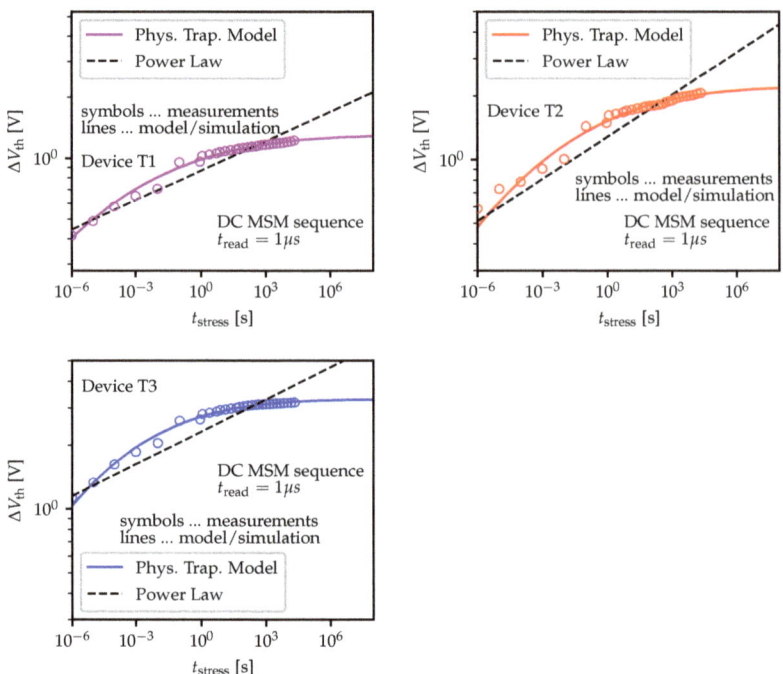

Figure 2. The prediction of the two-state model is shown versus power-law like functions for the three technologies evaluated in this work: T1 (**top-left**), T2 (**top-right**) and T3 (**bottom**). As can be seen, our simulations nicely replicate the experimental data, while the power-law (black) significantly deviates from the experimental data. The measurement delay for recording V_G after stress for this setup is $t_{read} = 1\,\mu s$.

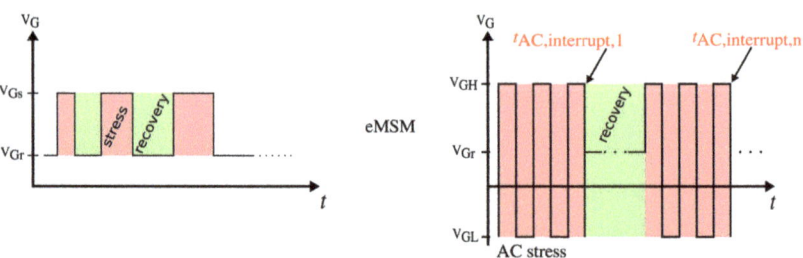

Figure 3. Schematic of the MSM measurement sequences of the input signals that are applied to the gate terminal of the transistors. The ΔV_{th} values are extracted considering, (**left**) a long-term DC PBTI degradation where a constant stress bias is applied before the recovery phase is recorded, and (**right**) a short-term AC ΔV_{th} is measured at different time points ($t_{AC,interrupt}$) at which the short AC stress is interrupted [22].

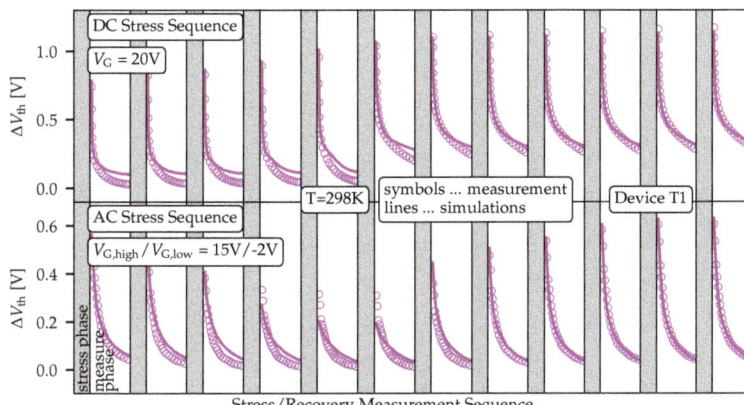

Figure 4. Comparison between simulation and experimentally ΔV_{th} extracted values for T1. The simulations nicely reproduce the measured data set ΔV_{th}.

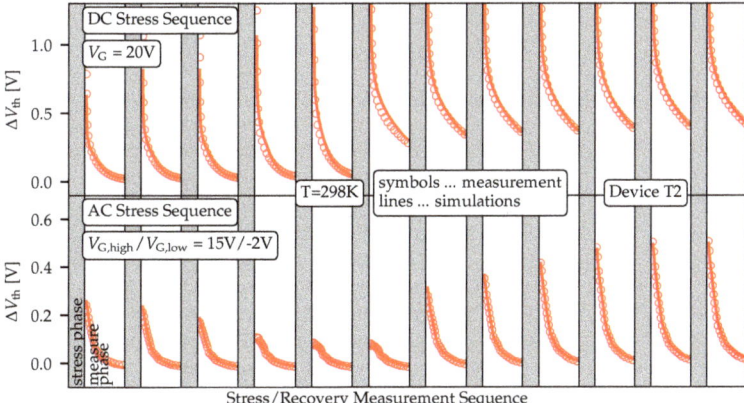

Figure 5. Similar measurement sequence as shown in Figure 4 is presented here for T2. Again our model agrees well with the experimental data.

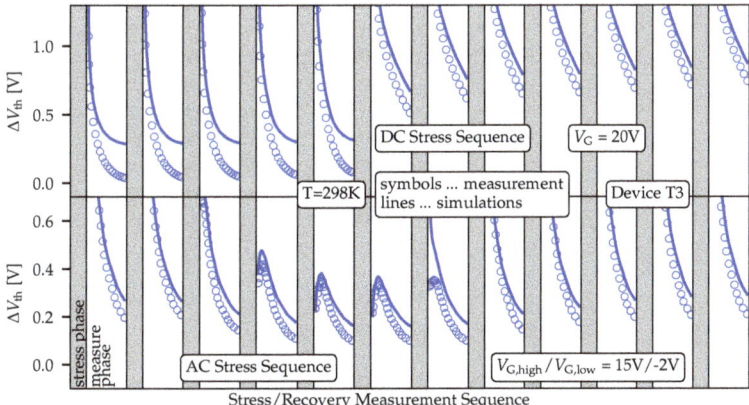

Figure 6. Also the data recorded employing T3 can be reconstructed at high accuracy by our physical computer models.

3. Compact Modeling of BTI for Circuit Simulations

As has been previously discussed, with the compact models provided for circuit simulations, the static device behavior can be nicely reproduced. However, such models do not account for time-dependent changes in the device characteristics. To consider the impact of BTI on the transistors using spice simulations, we have to extend the model by adding an independent voltage source to the gate of the transistors. This additional voltage source accounts for the threshold voltage drift, ΔV_{th} [23], and the schematic of the modified model is depicted in Figure 7.

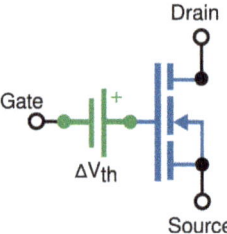

Figure 7. The BTI impact on the transistors is represented by adding a voltage source with a value equal to ΔV_{th} to the gate of the devices implemented in the Spice simulator. ΔV_{th} values are extracted from the accurate physical defect trapping model.

Since the transistors are typically used in switching applications, transient circuit simulations are carried out next. For this purpose, we perform AC simulations employing *ngspice* and combine them with calculations made using our reliability tool for accounting the ΔV_{th}. The additional voltage source connected to the gate of the transistor contains the variability/degradation effects predicted employing the physical defect trapping model [24] implemented in the reliability simulator Comphy [21].

With each simulation, considering the accurate ΔV_{th} values, it is possible to demonstrate that an overestimation of the ΔV_{th} extracted values using power-law-like functions. Likewise, this difference in the ΔV_{th} can lead to a very pessimistic prediction of inverter circuits parameters such as the propagation delay time t_D.

In Figure 8, the simulation process flow using the open-source *ngspice* simulator is illustrated. To perform our reliability simulations, we define the net-list of the circuit based on the spice models of the respective SiC power MOSFETs (step 1). Next, an external file contains the ΔV_{th} values extracted from Comphy (step 2) for a long operation time (ten years). This file is used to modify the initial net-list of the circuit. Then we extract the aged parameters of the circuit for a specific degradation and time point, taking into account the principle shown in Figure 7 (step 3). Note that such simulations can become easily computationally demanding. The simulation time depends on the number of ΔV_{th} values in the external file. The *ngspice* simulator launches a new simulation every time that the initial net-list is modified. However, this net-list modification does not significantly increase the simulation time. As a result of the circuit simulations, we obtain the aged circuit parameters at extended operation time considering the impact of BTI on the SiC MOSFETs. We can process these results to analyze the plots of the aged electrical characteristics of the device (step 4).

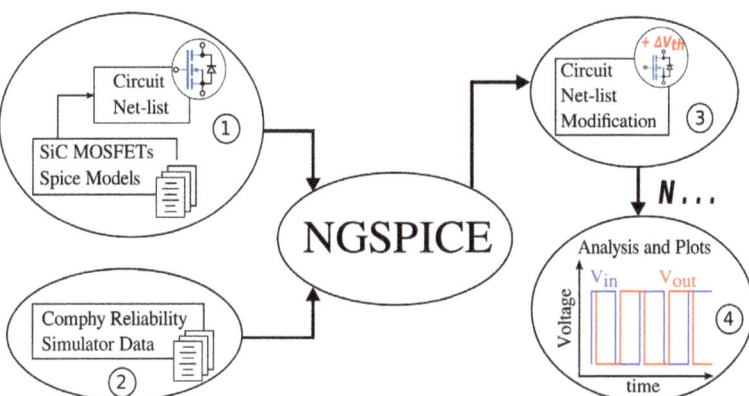

Figure 8. Simulation process flow using open-source simulator *ngspice*, to analyze the BTI impact on the inverter circuits performance. The spice simulator uses the data from the reliability simulator Comphy to modify the initial net-list of the circuit. It launches simulations to extract the aged circuit parameters for long operation time. Each simulation (step 3) corresponds to an inevitable degradation of the transistors for a specific time (step 2).

4. Results and Discussion

The first circuit we evaluate is the typical resistivity load inverter, see Figure 9 **(left)**. With this circuit, we can efficiently verify our approach and implementation to determine the BTI impact in the propagation delay t_D considering only one transistor. The values are extracted as the time difference when V_{in} and V_{out} are equal to $V_{DD}/2$, as we show in Figure 9 **(right)**. Each value of the delay time distribution t_D represents the aging of the inverter circuit for each ΔV_{th} or, equivalently, for a certain degradation of the device with time.

In Figure 10 shows the ΔV_{th} values extracted after ten years ($\approx 10^8$ s) of operational time considering an AC input signal with a typical switching frequency for SiC applications equal to f_{SW} = 50 kHz, duty cycle of 50 %, V_{High} = 20 V and V_{Low} = 0 V. The time of the zero voltage in the input signal ensures the recovery phase for the ΔV_{th} extraction. Likewise, we consider the case for $V_{High} = V_{Low}$ = 20 V (DC stress) to compare the ΔV_{th} values extracted when the recovery phase is omitted. For all devices the ΔV_{th} values are presented in the same plot for a long-term DC and short-term AC stress sequences to highlight the differences obtained between Comphy simulations with and without recovery. As can be seen, the latter leads to a significant overestimation of the ΔV_{th}.

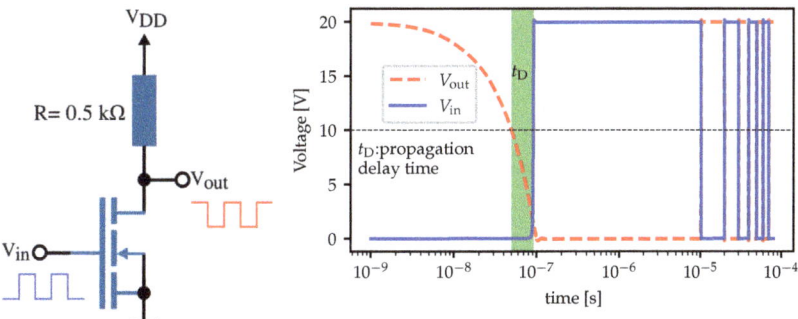

Figure 9. Schematic of the resistivity load inverter used as simple test circuit with $R = 0.5$ kΩ, V_{DD} = 20 V **(left)** and propagation delay time t_D extraction at T = 298 K **(right)**. The latter is defined as the time difference when the output V_{out} and input voltage V_{in} equals $V_{DD}/2$.

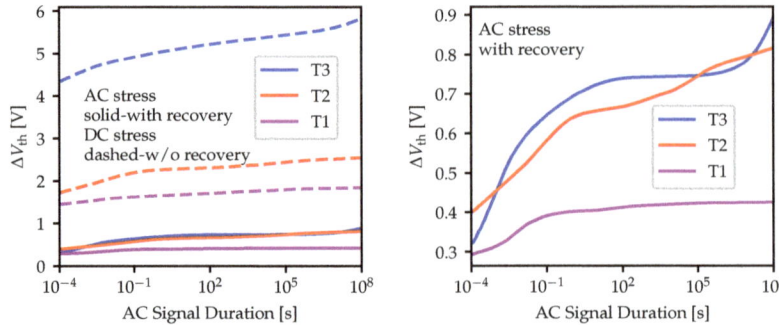

Figure 10. Threshold voltage drift (ΔV_{th}) behavior extracted employing Comphy for the different SiC power MOSFETs used in this work, i.e., T1, T2 and T3 at room temperature (T = 298 K), for an AC input signal with duty cycle = 0.5, f = 50 kHz and a stress time equal to t_{str} = 100 ms. A comparison of the extracted values when the recovery phase is considered and omitted is shown (**left**). A more detailed representation of the short-term AC stress case is depicted (**right**).

As can be seen, the threshold voltage of the transistors drifts around 0.3–0.9 V after ten years of operation if the recovery phase is considered. Almost five times more, between 1.5–6 V, if the recovery phase is omitted. The propagation delay time for the resistivity load inverter circuit based on the different commercial SiC transistors T1, T2, and T3, considering the ΔV_{th} extracted values, are presented in Figure 11. These results represent the BTI impact on the simple test inverter circuit considering only the degradation of one transistor independently for a long operating time. Considering ΔV_{th} = 0 (fresh simulations), the extracted propagation delay time for the three technologies was 39 ns, 78 ns, and 169 ns for T1, T2, and T3, respectively. An increase of the threshold voltage drift over time also causes an increase in the propagation delay time of the inverter circuit. In all the cases, the increase of the propagation delay time Δt_D is expressed in (%) and represents the difference between the t_D for a specific operating time ($\Delta V_{th} \neq 0$) and the initial value (ΔV_{th} = 0). After ten years of operating time t_D increases up to 41 ns, 85 ns, and 172 ns, for T1, T2, and T3, respectively. From these results, we can note that the increase of the propagation delay time is similar (between 2–4 ns) for each technology. However, the BTI impact (computed as an increase of the propagation delay time, see Figure 11) on T1/T2 technologies (8 %) is more significant than for T3 (1.5 %). One possible explanation for this is due the propagation delay time is one order higher for T3 technology. Finally, our results also show that when the recovery phase is omitted, the t_D is overestimated for all technologies.

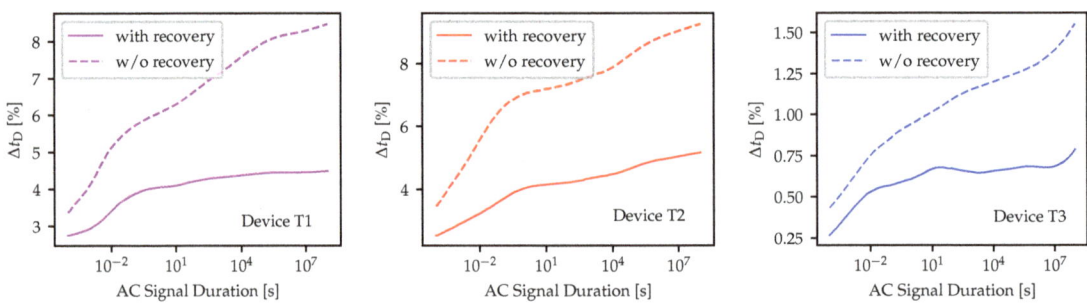

Figure 11. The impact of BTI on the propagation delay time for the resistivity load inverter circuit based on T1 (**left**), T2 (**center**), and T3 devices (**right**) is shown. For the three technologies, after ten years of the operating time, t_D is overestimated around by 2–4 ns if the recovery phase in the AC input signal is omitted. The results show a BTI impact less significant on T3 than in the other ones.

The following circuit we analyze is the pseudo-D CMOS inverter. In literature, this circuit is proposed for thin-film transistors [25,26] to implement inverters employing mono-type transistors. We make use of our simulation approach and extract the t_D for a pseudo-D CMOS inverter circuit, shown in Figure 12, designed with mono-type transistors [27]. For the analysis of this circuit, it is essential to mention that we only use the T1 device. Considering that these devices were fabricated for only one channel width and length, it was impossible to design this inverter with T2 and T3 due to impedance coupling. However, thanks to the Kelvin Source pin architecture (which allows more voltage applied between the gate and source, resulting in a faster dynamic switching and reduction of the inductive effects at the gate of the device), it was possible to design and simulate the pseudo-D CMOS inverter circuit using the T1 device.

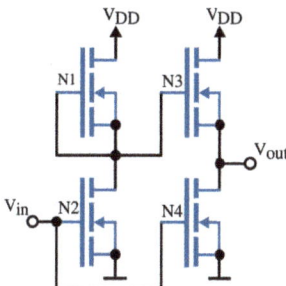

Figure 12. Schematic representation of the pseudo-D CMOS inverter circuit using commercially available SiC power MOSFETs. We analyze the BTI impact on the circuit performance considering the degradation of the V_{th} of the transistors.

For this circuit, we compute the total degradation by evaluating ΔV_{th} of each transistor individually at each simulation time step. Figure 13 shows the t_D distribution values for ten years of operation considering the degradation of all transistors at the same time and the degradation of each transistor individually. An overestimation of the propagation delay of the pseudo-D logic inverter circuit for a long operation time can be observed. The propagation delay t_D is a few nanoseconds larger when the recovery phase of the AC signal is omitted for each transistor. This can lead to challenges for circuits operating at higher frequencies. Note that this difference in the predicted delay does not arise from real devices, but rather is a consequence of the inaccuracy of simple models. The results for N2 and N4 show a similar impact on t_D. On the other hand, for N1 and N3, the circuit seems to be practically insensitive to BTI.

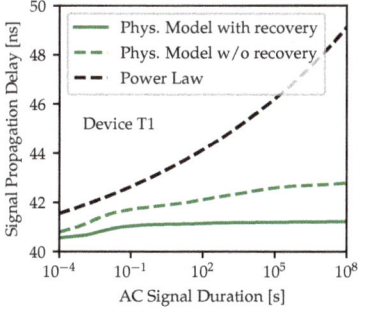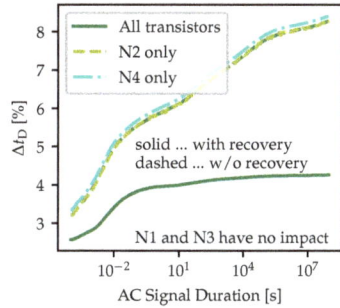

Figure 13. Extracted signal propagation delay for the pseudo-CMOS inverter circuit considering the impact of BTI for ten years of operating time. A comparison of the results when simple power-law-like functions and physical models are used is shown (**left**). While N1 and N3 have no to impact on t_D, the propagation delay is affected by N2 and N4 in a similar way (**right**).

5. Conclusions

For a highly accurate description of the behavior of aging circuits, physical device models are required for circuit simulations. However, existing approaches use compact models that are typically based on simple mathematical formulas for circuit analysis and cannot account for aging mechanisms such as BTI at a very high accuracy over a wide range of temperature, bias and operating conditions. In our work, we combine our transistor reliability simulator Comphy with circuit simulations made using *ngspice*. We employ three transistor technologies and extend the compact models provided by the device vendors to be suitable for precise BTI evaluation. After Comphy is calibrated to extensive DC and AC measurements, we evaluate the propagation delay of the inverter circuits. Specifically, we analyze the resistivity load inverter circuit and the pseudo-D-CMOS inverter based on the same three different technologies, T1, T2, and T3. Our results show that the BTI has a more significant impact on T1/T2 technology than T3. On the other hand, the same results demonstrate that employing simple power-law-like formulas leads to an overestimation of the signal propagation delay, leading to severe challenges during circuit design. We also demonstrate that omitting device recovery during the low-phase under AC operation leads to a considerable overestimation of the signal propagation delay of the circuits.

Author Contributions: Conceptualization, Y.H. and M.W.; methodology, Y.H., B.S., T.G. and M.W.; investigation, Y.H.; writing—review and editing, Y.H., B.S., T.G. and M.W.; supervision, T.G. and M.W. All authors have read and agreed to the published version of the manuscript.

Funding: The financial support by the Austrian Federal Ministry for Digital and Economic Affairs, the National Foundation for Research, Technology and Development and the Christian Doppler Research Association is gratefully acknowledged. Furthermore, The research leading to these results has received funding from the from the Take-off program of the Austrian Research Promotion Agency FFG (projects no. 867414, and 861022).

Acknowledgments: Stimulating discussions with and support by Katja Waschneck and Hans Reisinger, both Infineon Munich, are gratefully acknowledged. The authors acknowledge TU Wien Bibliothek for financial support through its Open Access Funding Programme.

Conflicts of Interest: The authors declare no conflict of interest.

References

1. Pulvirenti, M.; Montoro, G.; Nania, M.; Scollo, R.; Scelba, G.; Cacciato, M.; Scarcella, G.; Salvo, L. *2018 IEEE Energy Conversion Congress and Exposition (ECCE)*; IEEE: Portland, OR, USA, 2018; pp. 1895–1902.
2. Dimitrijev, S. SiC Power MOSFETs: The current status and the potential for future development. In Proceedings of the 30th International Conference on Microelectronics (MIEL 2017), Nis, Serbia, 9–11 October 2017. [CrossRef]
3. Litchtenwalner, D.J.; Hull, B.; Van Brunt, E.; Sabri, S.; Gajewski, D.A.; Grider, D.; Allen, S.; Palmour, J.W. Reliability studies of SiC vertical power MOSFETs. In Proceedings of the 2018 IEEE International Reliability Physics Symposium (IRPS), Burlingame, CA, USA, 11–15 March 2018. [CrossRef]
4. Berens, J.; Weger, M.; Pobegen, G.; Aichinger, T.; Rescher, G.; Schleich, C.; Grasser, T. Similarities and Differences of BTI in SiC and Si Power MOSFETs. In Proceedings of the IRPS, Dallas, TX, USA, 28 April–30 May 2020. [CrossRef]
5. Barbagallo, C.; Rizzo, S.; Scelba, G.; Scarcella, G.; Cacciato, M. On the Lifetime estimation of SiC power MOSFETs for motor drive applications. *Electronics* **2021**, *10*, 324. [CrossRef]
6. Guevara, E.; Herrera-Perez, V.; Rocha, C.; Guerrero, K. Threshold Voltage Degradation for n-Channel 4H-SiC power MOSFETs. *J. Low Power Electron. Appl.* **2020**, *10*, 3. [CrossRef]
7. www.wolfspeed.com. Available online: https://www.wolfspeed.com/document-library/?documentType=ltspice-and-plecs-models&productLine=power (accessed on 10 July 2021).
8. www.st.com. Available online: https://www.st.com/en/power-transistors/wide-bandgap-transistors.html#cad-resource (accessed on 10 July 2021).
9. Waltl, M.; Rzepa, G.; Grill, A.; Wolfgang, G.; Franco, J.; Kaczer, B.; Witters, L.; Mitard, J.; Horiguchi, N.; Grasser, T. Superior NBTI in High-k Si-Ge Transistors—Part I: Experimental. *IEEE Trans. Electron Devices* **2017**, *64*, 2092–2098. [CrossRef]
10. Habersat, D.B.; Lelis, A.J.; Green, R. Measurement considerations for evaluating BTI effects in SiC MOSFETs. *Microelectron. Reliab.* **2018**, *81*, 121–126. [CrossRef]
11. Green, R.; Lelis, A.; Habersat, D. Threshold-voltage bias-temperature instability in commercially-available SiC MOSFETs. *Jpn. J. Appl. Phys.* **2016**, *55*. [CrossRef]

12. Peters, D. Investigation of threshold voltage stability of SiC MOSFETs. In Proceedings of the 2018 IEEE 30th International Symposium on Power Semiconductor Devices and ICs (ISPSD), Chicago, IL, USA, 13–17 May 2018; pp. 42–43. [CrossRef]
13. Toledano-Luque, M.; Kaczer, B.; Simoen, E.; Roussel, P.; Veloso, A.; Grasser, T.; Groeseneken, G. Temperature and Voltage Dependences of the Capture and Emission Times of Individual Traps in High-k Dielectrics. *Microelectron. Eng.* **2011**, *88*, 1243–1246. [CrossRef]
14. Kim, J.J.; Linder, B.P.; Rao, R.M.; Kim, T.H.; Lu, P.F.; Jenkins, K.A.; Kim, C.H.; Bansal, A.; Mukhopadhyay, S.; Chuang, C.T. Reliability monitoring ring oscillator structures for isolated/-combined NBTI and PBTI measurement in high-k metal gate technologies. In Proceedings of the 2011 International Reliability Physics Symposium, Monterey, CA, USA, 10–14 April 2011. [CrossRef]
15. Miki, H.; Tega, N.; Yamaoka, M.; Frank, D.J.; Bansal, A.; Kobayashi, M.; Cheng, K.; D'Emic, C.P.; Ren, Z.; Wu, S.; et al. Statistical measurement of random telegraph noise and its impact in scaled-down high-k/metal-gate MOSFETs. In Proceedings of the 2012 International Electron Devices Meeting, San Francisco, CA, USA, 10–13 December 2012. [CrossRef]
16. Rzepa, G.; Franco, J.; Subirats, A.; Jech, M.; Chasin, A.; Grill, A.; Waltl, M.; Knobloch, T.; Stampfer, B.; Chiarella, T.; et al. Efficient Physical Defect Model Applied to PBTI in High-k Stacks. In Proceedings of the 2017 IEEE International Reliability Physics Symposium (IRPS), Monterey, CA, USA, 2–6 April 2017. [CrossRef]
17. Schleich, C.; Waldhoer, D.; Waschneck, K.; Feil, M.W.; Reisinger, H.; Grasser, T.; Waltl, M. Physical Modeling of Charge Trapping in 4H-SiC DMOSFET Technologies. *IEEE Trans. Electron Devices* **2021**, *68*, 4016–4021. [CrossRef]
18. Lelis, A.J.; Habersat, D.; Green, R.; Ogunniyi, A.; Gurfinkel, M.; Suehle, J.; Goldsman, N. Time Dependence of Bias-Stress-induced SiC MOSFET Threshold-Voltage Instability Measurements. *IEEE Trans. Electron Devices* **2008**, *55*, 1835–1840. [CrossRef]
19. Schleich, C.; Berens, J.; Rzepa, G.; Pobegen, G.; Tyaginov, S.; Grasser, T.; Waltl, M. Physical Modeling of Bias Temperature Instabilities in SiC MOSFETs. In Proceedings of the 2019 IEEE International Electron Devices Meeting (IEDM), San Francisco, CA, USA, 7–11 December 2019. [CrossRef]
20. Grasser, T. Stochastic Charge Trapping in Oxides: From Random Telegraph Noise to Bias Temperature Instabilities. *Microelectron. Reliab.* **2012**, *52*, 39–70. [CrossRef]
21. Rzepa, G.; Franco, J.; O'Sullivan, B.; Subirats, A.; Simicic, M.; Hellings, G.; Weckx, P.; Jech, M.; Knobloch, T.; Waltl, M.; et al. Comphy—A compact-physics framework for unified modeling of BTI. *Microelectron. Reliab.* **2018**, *85*, 49–65. [CrossRef]
22. Puschkarsky, K.; Grasser, T.; Aichinger, T.; Gustin, W.; Reisinger, H. Review on SiC MOSFETs High-Voltage Device Reliability Focusing on Threshold Voltage Instability. *IEEE Trans. Electron Devices* **2019**, *66*, 4604–4616. [CrossRef]
23. Martin-Martinez, J.; Ayala, N.; Rodriguez, R.; Nafria, M.; Aymerich, X. RELAB: A tool to include MOSFETs BTI and variability in SPICE simulators. In Proceedings of the 2012 International Conference on Synthesis, Modeling, Analysis and Simulation Methods and Applications to Circuit Design (SMACD), Seville, Spain, 19–21 September 2012. [CrossRef]
24. Grasser, T.; Kaczer, B.; Goes, W.; Reisinger, H.; Aichinger, T.; Hehenberger, P.; Wagner, P.J.; Schanovsky, F.; Franco, J.; Roussel, P.; et al. Recent Advances in Understanding the Bias Temperature Instability. In Proceedings of the 2010 International Electron Devices Meeting, San Francisco, CA, USA, 6–8 December 2010. [CrossRef]
25. Fukuda, K.; Sekitani, T.; Yokota, T.; Kuribara, K.; Huang, T.C.; Sakurai, T.; Zschieschang, U.; Klauk, H.; Ikeda, M.; Kuwabara, H.; et al. Organic Pseudo-CMOS Circuits for Low-Voltage Large-Gain High-Speed Operation. *IEEE Electron Device Lett.* **2011**, *32*, 1448–1450. [CrossRef]
26. Huang, T.C.; Fukuda, K.; Lo, C.M.; Yeh, Y.H.; Sekitani, T.; Someya, T.; Cheng, K.T. Pseudo-CMOS: A Design Style for Low-Cost and Robust Flexible Electronics. *IEEE Trans. Electron Devices* **2011**, *58*, 141–150. [CrossRef]
27. Kuroki, S.; Kurose, T.; Nagatsuma, H.; Ishikawa, S.; Maeda, T.; Sezaki, H.; Kikkawa, T.; Makino, T.; Ohshima, T.; Östling, M.; et al. 4H-SiC Pseudo-CMOS Logic Inverters for Harsh Environment Electronics. *Mater. Sci. Forum* **2017**, *897*, 669–672. [CrossRef]

Article

Photo-Excited Switchable Terahertz Metamaterial Polarization Converter/Absorber

Dingwang Yu [1,2], Yanfei Dong [1,*], Youde Ruan [1], Guochao Li [1], Gaosheng Li [2], Haomin Ma [1], Song Deng [1] and Zhenpeng Liu [1]

1. Industries Training Centre, Shenzhen Polytechnic (SZPT), Shenzhen 518055, China; ydw13049059@163.com (D.Y.); ruyude01@szpt.edu.cn (Y.R.); liguochao@szpt.edu.cn (G.L.); mahaomin@szpt.edu.cn (H.M.); ds1210@szpt.edu.cn (S.D.); lzp790310@szpt.edu.cn (Z.L.)
2. College of Electrical and Information Engineering, Hunan University, Changsha 410082, China; gfkdlgs@163.com
* Correspondence: dongyanfei@szpt.edu.cn

Citation: Yu, D.; Dong, Y.; Ruan, Y.; Li, G.; Li, G.; Ma, H.; Deng, S.; Liu, Z. Photo-Excited Switchable Terahertz Metamaterial Polarization Converter/Absorber. *Crystals* 2021, 11, 1116. https://doi.org/10.3390/cryst11091116

Academic Editor: Michael Waltl

Received: 9 August 2021
Accepted: 10 September 2021
Published: 14 September 2021

Publisher's Note: MDPI stays neutral with regard to jurisdictional claims in published maps and institutional affiliations.

Copyright: © 2021 by the authors. Licensee MDPI, Basel, Switzerland. This article is an open access article distributed under the terms and conditions of the Creative Commons Attribution (CC BY) license (https://creativecommons.org/licenses/by/4.0/).

Abstract: In this paper, a photo-excited switchable terahertz metamaterial (MM) polarization converter/absorber has been presented. The switchable structure comprises an orthogonal double split-ring resonator (ODSRR) and a metallic ground, separated by a dielectric spacer. The gaps of ODSRR are filled with semiconductor photoconductive silicon (Si), whose conductivity can be dynamically tuned by the incident pump beam with different power. From the simulated results, it can be observed that the proposed structure implements a wide polarization-conversion band in 2.01–2.56 THz with the conversion ratio of more than 90% and no pump beam power incident illuminating the structure, whereas two absorption peaks operate at 1.98 THz and 3.24 THz with the absorption rates of 70.5% and 94.2%, respectively, in the case of the maximum pump power. Equivalent circuit models are constructed for absorption states to provide physical insight into their operation. Meanwhile, the surface current distributions are also illustrated to explain the working principle. The simulated results show that this design has the advantage of the switchable performance afforded by semiconductor photoconductive Si, creating a path towards THz imaging, active switcher, etc.

Keywords: metamaterial polarization converter/absorber; switcher; photoconductive silicon; THz wave; orthogonal double split-ring resonator

1. Introduction

Metamaterials (MMs), a class of artificial materials comprised of sub-wavelength periodic or non-periodic structures, have received more and more interest in the past few decades [1] due to their extraordinary characteristics in manipulating electromagnetic (EM) waves unavailable in nature [2]. MMs have been widely used in a variety of functional devices so far, such as absorber [3–8], cloak [9–12], sensor [13–15], polarizer [16–19] and so on.

With the development of terahertz (THz) techniques and materials, various MM-based THz devices have been deployed and designed to manipulate THz waves over the past few decades [20–24]. Among these architectures, in general, the MM structures can be divided into two types: reflection [21,22] and transmission [23,24]. However, most THz devices usually can only work in static (reflection/transmission state), and thus have a single function making them difficult to change once fabricated, which severely hamper their practical applications.

To solve this challenge, MM structures integrated with active media (i.e., active THz devices), such as MEMS [25,26], graphene [27–29], vanadium dioxide (VO$_2$) [30,31], indium antimonide (InSb) [32,33] and semiconductors silicon (Si) [34,35], etc., have been presented and designed to realize the dynamic and active manipulation of THz wave under the control of external stimuli, such as electrical biasing, optical illumination and thermal excitation.

Among these tunable materials, the photoconductive semiconductors (Si) [36,37] can provide a viable pathway to realize a fast change of the reflection/transmission responses for the incident waves under the excitation of light pulse with its exceptional optical-electrical characteristics, including ultrafast response, low cost and high quantum efficiency.

Recently, several active MM-based THz structures based on photoconductive silicon have been investigated in-depth with a lot of effort [38–41]. These designs can demonstrate unique advantages, including the operational state, the working frequency and intensity modulation to ensure a fair comparison with existing devices at hand, but these architectures are incapable of realizing the switchable performance between absorption [42,43] and polarization conversion [44–46], and it only can work for a single function.

In this paper, a photo-excited switchable MM polarization converter/absorber is proposed with photoconductive silicon (Si), which can be freely and continuously switched from a broadband polarization converter to a dual-band absorber in two different pump beam power. Due to the conductivity of photoconductive Si (σ_{Si}) being proportional to the pump power of the incident optical beam, the switchable capability of this structure can be achieved by dynamically adjusting the working state of photoconductive Si. This structure can cover a 24.1% fractional bandwidth of polarization conversion with the polarization-conversion ratio (PCR) >90% as σ_{Si} = 1 S/m, while σ_{Si} is equal to 1×10^5 S/m, where it can behave as a dual-band absorber. The surface current distributions of this design on both top and bottom layers are provided to investigate the switchable operating mechanism for different Si conductivity states. This structure would be poised to act as a suitable alternative to THz sensing, communication and detection, etc., for its excellent characteristics.

2. Metamaterials and Methods
2.1. Metamaterials Model

The unit cell geometry of the proposed structure is shown in Figure 1. From the figure, the structure consists of a top metallic orthogonal double split-ring resonator (ODSRR) and a dielectric substrate with a bottom ground plane. The gold is selected as a metallic model for this structure with a thickness (t) of 0.4 µm, and conductivity (σ) of 4.561×10^7 S/m. The dielectric layer is polyimide material ($\varepsilon_r = 3.5$, $\tan \delta = 0.02$) with a thickness (t_s) of 6.5 µm. The semiconductor photoconductive Si (blue part) is integrated into the split gaps of ODSRR, which can be modeled as a dielectric material ($\varepsilon_{Si} = 11.7$) with a thickness (t) of 0.4 µm, whose conductivity (σ_{Si}) changes with variation of the incident pump beam power. Then other geometric parameters of the proposed structure (µm) are $a = 30$, $r_1 = 13.5$, $r_2 = 10.5$, $w = 2$, $g = 0.5$.

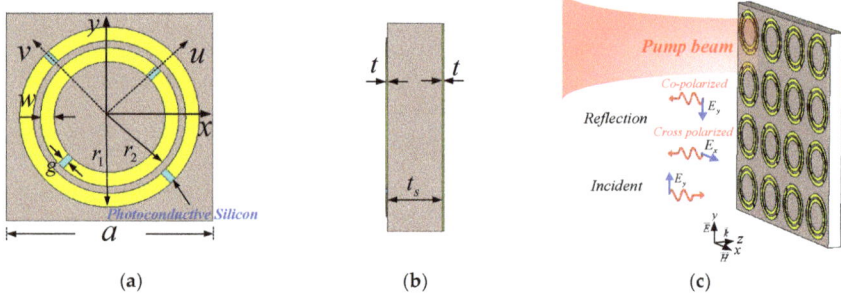

Figure 1. Schemes of the proposed switchable MM polarization converter/absorber structure: (**a**) The front and (**b**) side views of the unit cell structure; (**c**) 2-D array.

2.2. Mathematical Method

To better explain the switchable property of the proposed structure, the uv coordinate system is introduced to mark the anisotropic axes, and both u and v axes exhibit 45° phase shifts as compared to the x and y axes, respectively, as shown in Figure 1a. To effectively analyze the polarization characteristic of the polarization converter, the co-polarization and cross-polarized reflections can be defined as $r_{yy} = |E_{ry}/E_{iy}|$ and $r_{xy} = |E_{rx}/E_{iy}|$ for the y-polarized incident wave [47], where the subscripts of i and r represent the incident and reflected wave modes, respectively, and then the subscripts of x and y indicate the electric field directions. The phase difference between the y and x components of the reflected THz wave is also written as $\Delta\phi = \phi_{xy} - \phi_{yy}$. To estimate the polarization conversion performance, the polarization rotation azimuth angle φ and the polarization-conversion ratio (PCR) can be extracted from the refection coefficients [48] to be targeted as goal metrics during design. Therefore, φ can be calculated as:

$$\varphi = \frac{1}{2}\text{acr}\tan[\frac{2R\cos(\Delta\phi)}{1-R^2}] \tag{1}$$

where $R = |r_{xy}|/|r_{yy}|$ and PCR can be obtained in the following manner:

$$PCR = \frac{|r_{xy}|^2}{|r_{xy}|^2 + |r_{yy}|^2} \tag{2}$$

As the bottom layer is a metallic plane, the transmission is nearly zero and, thus, the absorptivity of this design can be defined as:

$$A = 1 - R = 1 - (|r_{xy}|^2 + |r_{yy}|^2) \tag{3}$$

3. Results and Discussions

To demonstrate the switchable performance of this structure, the numerical model is constructed to simulate with the commercial full-wave solver, software CST Microwave Studio, for two different Si conductivity (σ_{Si}) states. In the simulation setting, the periodic boundary conditions (PBC) oriented along the x and y directions is used to model the periodic structure with a normal wave incident upon the unit cell with the E-field vector in the y axis, as described in detail in Figure 1c, behaving as the exciting source.

3.1. Reflection Responses

The simulated reflection responses as a function of frequency for two different conductivity states are illustrated in Figures 2 and 3. In the case of $\sigma_{Si} = 1\,\text{S/m}$ without pump beam power, the cross-polarization r_{xy} is much greater than the co-polarization r_{yy} across the operating band of 1.8–2.7 THz as plotted in Figure 2a. In Figure 2b, it can be seen that PCR is more than 0.9 in the frequency range of 2.01–2.56 THz with an absorption rate less than 0.3. Meanwhile, the rotation azimuth angle is approximately around ±90° in this band, forming a broad cross-polarization conversion bandwidth. Hence, for this case, the designed structure can be referred to as a broad polarization converter.

With the maximal pump beam power incident on the structure, σ_{Si} can reach up to $1 \times 10^5\,\text{S/m}$, termed mental state, such that the Si-filled gaps would be in short circuit state, then the cross-polarization r_{xy} is less than the co-polarization r_{yy} over the whole frequency band as observed from Figure 3a. From the results in Figure 3b, the PCR is below 0.2 at the two resonant peaks of 1.98 THz and 3.24 THz, respectively, and the corresponding absorption rates are around 70.5% and 94.2%, respectively, with the rotation azimuth angles less than 20° across the whole frequency band. Thereby, the structure could be used as a dual-band absorber. Thus, this proposed hybrid metal-semiconductor ODSRR structure could be switched to a polarization converter or absorber by the semiconductor photoconductive Si which can act as the active THz component with different working states under different external pumps' beam power.

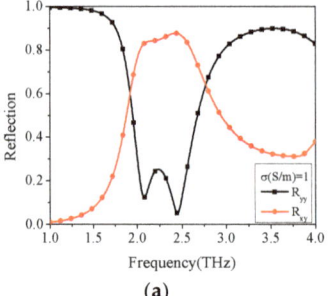 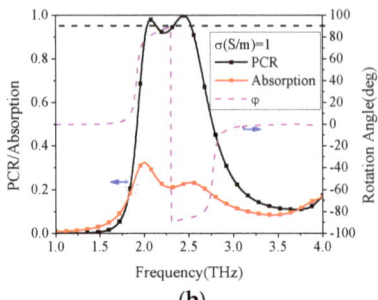

Figure 2. Simulated (**a**) cross- and co-polarization reflection coefficients and (**b**) PCR, absorption and rotation angle with $\sigma_{Si} = 1$ S/m.

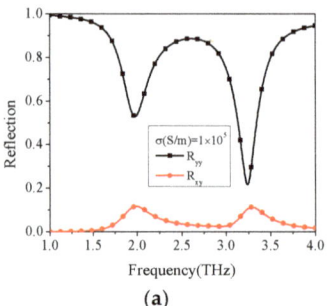 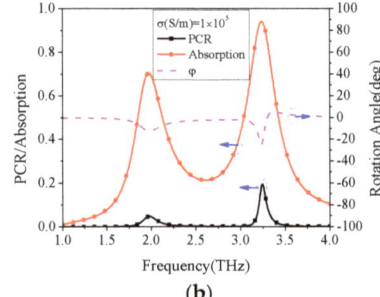

Figure 3. Simulated (**a**) cross- and co-polarization reflection coefficients and (**b**) PCR, absorption and rotation angle with $\sigma_{Si} = 1 \times 10^5$ S/m.

3.2. Validation of the Equivalent Circuit Model

In an attempt to analytically describe the absorption operation, the schematic description of the equivalent circuit model (ECM) for this structure is shown in Figure 4a. The double metallic rings can be represented by distributive elements, whereas the substrate is considered as a transmission line with the length of t_s and the wave impedance $Z_s = Z_0/\sqrt{\varepsilon_r}$, Z_0 is the characteristic impedance of the free space. C_m represents the electrical coupling between two double-opening coupling rings [49]. The values of the reactive elements can be approximately calculated as [50,51]:

$$C_{i,o} = \varepsilon_0 \varepsilon_{eff} \frac{2a}{\pi} \ln(\csc \frac{(a-r_{1,2})\pi}{2a}) \tag{4}$$

$$L_{i,o} = \mu_0 \mu_{eff} \frac{a}{2\pi} \ln(\csc \frac{w\pi}{2a}) \tag{5}$$

where ε_0 and μ_0 are the permittivity and permeability of free space, respectively. ε_{eff} and μ_{eff} denote the effective permittivity and permeability of the supporting substrate, respectively. The series circuits *RLC* provide the two absorption responses at 1.98 and 3.24 THz, respectively.

Then, the impedance of the top ODSRR surface can be indicated by Z_F, which is in parallel with Z_s. Therefore, the input impedance and reflection coefficient from this designed absorber can be respectively calculated as:

$$Z_{in} = Z_F || jZ_s \tan(\beta t_s) \qquad (6)$$

$$S_{11} = 20\log(\frac{Z_{in} - Z_o}{Z_{in} + Z_o}) \qquad (7)$$

To better validate the availability of ECM, the reflection characteristics calculated from the full-wave simulation in CST and the circuit model have been achieved for comparison below in Figure 4b, where good agreement can be seen between the two methods, sufficient to indicate the fact that the ECM used for the modeling method is valid and that results from the mathematical simulations constitute good predictions.

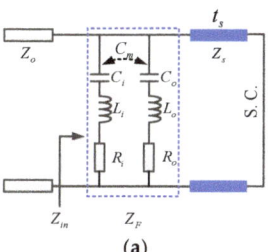
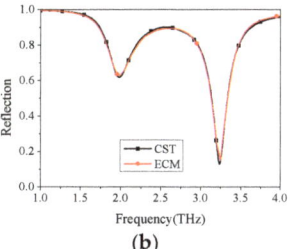

(a) (b)

Figure 4. (a) Equivalent transmission-line model of the proposed absorber (circuit parameters: C_i = 0.44 fF, C_o = 0.12 fF, C_m = 0.26 fF, L_i = 4.42 pH, L_o = 3.04 pH, R_i = 4.60 Ω, R_o = 4.78 Ω); (b) Comparison of reflectivity of the proposed structure calculated by the ECM and simulated in the CST software.

3.3. The Intrinsic Operation Mechanism

Meanwhile, to gain some insight on the working principle of switchable operation of this architecture, the surface current distributions on both top and bottom layers as $\sigma_{Si} = 1$ S/m and $\sigma_{Si} = 1 \times 10^5$ S/m under normal incidence are plotted in Figures 5 and 6, respectively, at four different frequencies, in which the arrows represent the direction of current flow and the color corresponds with the intensity.

As σ_{Si} = 1 S/m, the surface current distributions at the frequencies of 2.08 and 2.45 THz are described in Figure 5. For y-polarized incident EM wave, the induced surface currents at the top and bottom layers are in the anti-parallel direction, thus forming a circulating loop and exciting a magnetic resonance along the u-direction at 2.08 THz, which can generate the in-phase reflection E_{iv}, but instead E_{iu} is an out-of-phase reflection due to no v-direction magnetic resonance occurring. Hence, the $-90°$ polarization rotation will be implemented, and then the polarization direction of reflection response is converted from y- to x-axis at the resonant frequency. Similarly, the magnetic resonance operates at 2.45 THz with the E-field oriented along the v direction, providing the out-of-phase and in-phase reflections for E_{iv} and E_{iu}, respectively. Therefore, the y-to-x polarized reflection will be realized with $90°$ rotation.

As $\sigma_{Si} = 1 \times 10^5$ S/m for the maximal pump beam power case, the Si-filled gaps of the ODSRR structure are short-circuited since the semiconductor Si is in the conducting state. Thus, the ODSRR is treated as a double-ring resonator to lead to the high absorption performance. As observed from Figure 6, for the y-polarized incident wave, the surface currents mainly focus on the left and right sides of the outer ring at 1.98 THz, and at the frequency of 3.24 THz, the surface currents are also mainly distributed at the left and right arms of the inner ring. All these two absorption responses have a similar current distribution with that of the conventional ring-shaped MA, so it is worth noting that the absorption responses are originated from the two arranged dipoles. Therefore, the proposed structure possesses the ability to conduct the switching state between the broadband polarization converter and dual-band absorber for two different states.

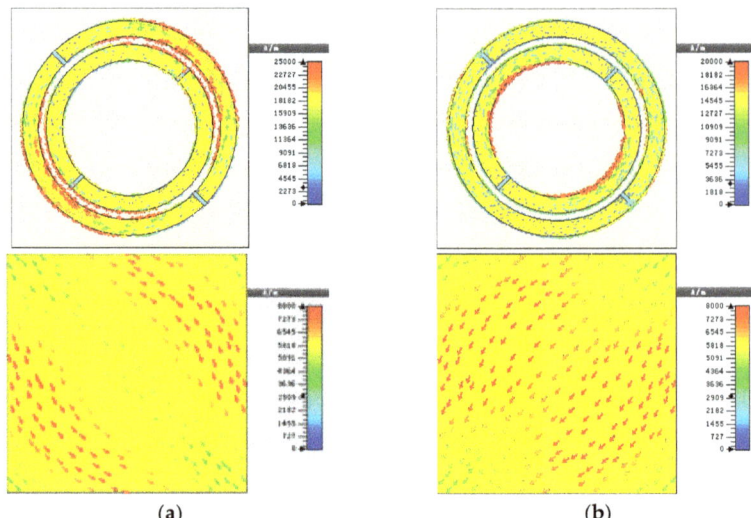

Figure 5. The surface current distributions on the top and bottom layers at (**a**) 2.08 and (**b**) 2.45 THz for $\sigma_{Si} = 1$ S/m.

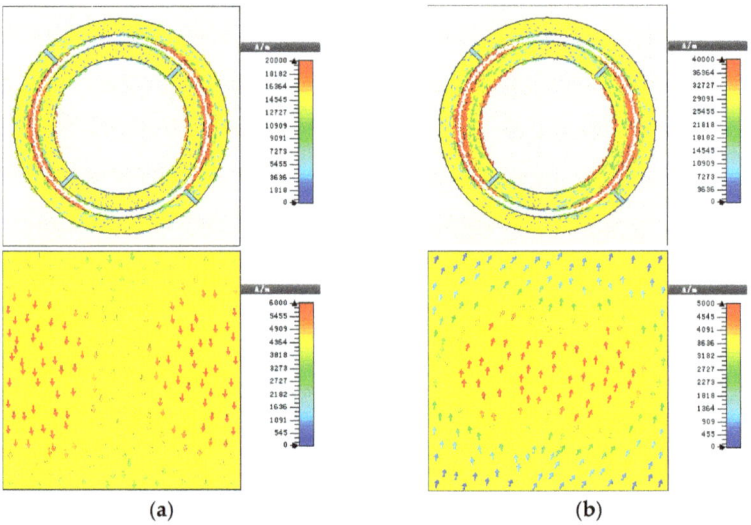

Figure 6. The surface current distributions on the top and bottom layers at (**a**) 1.98 and (**b**) 3.24 THz for $\sigma_{Si} = 1 \times 10^5$ S/m.

3.4. Oblique Incidence Characteristics

Figure 7 shows the oblique incidence characteristics for different Si conductivity (i.e., $\sigma_{Si} = 1$ S/m and $\sigma_{Si} = 1 \times 10^5$ S/m). From the results, it can be seen that the switchable structure maintains a wide operating bandwidth over the angle range from 0° to 45° with good PCRs of over 75% for both TE and TM waves in Figure 7a,b with $\sigma_{Si} = 1$ S/m. Figure 7c,d describe the absorption responses against incident angle (θ, the angle between the incident wave vector $\vec{k}$ and the z-axis) varying from 0° to 60° as $\sigma_{Si} = 1 \times 10^5$ S/m. In TE mode, the lower resonant frequency shifts slightly to the high frequency as θ goes up at 1.98 THz in Figure 7c, with the absorptivity gradually increasing. It can be ascribed to the strong electrical coupling between the outer and inner rings. On the contrary, the absorption performance is gradually deteriorated with θ changing at the upper frequency of 3.24 THz because the parallel *H*-field component decreases. For TM mode, the structure shows good angular stability when θ reaches up to 45° as detailed in Figure 7d. Though there is a slight frequency discrepancy (0.06 THz and 0.1 THz for TE and TM mode waves, respectively) for the lower absorption frequency, the upper absorption peak has better angular robustness than that of this absorption peak for different incidents' wave modes.

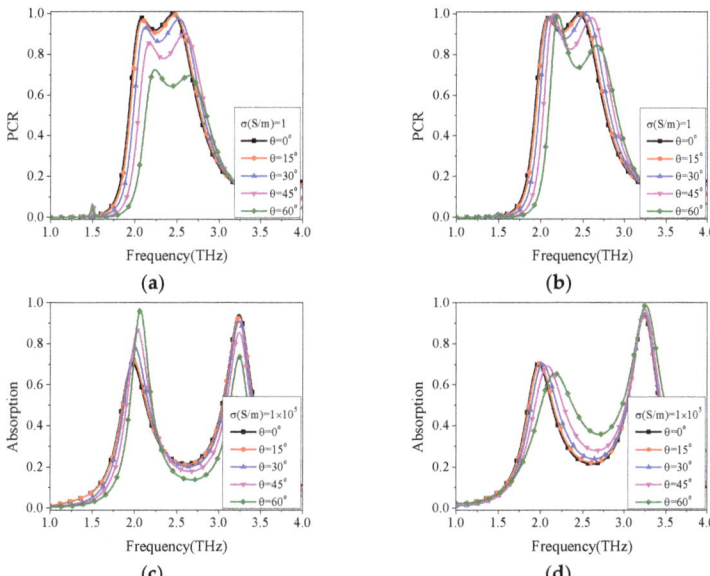

Figure 7. The PCR and absorption performances under different incident angles under (**a**,**c**) TE and (**b**,**d**) TM modes for different states.

A comparison with the current three materials, photoconductive Si, VO$_2$ and graphene embedded in structure to exhibit the switchable performance is illustrated in Figure 8. It is clearly apparent that the proposed design has achieved a better stable switching characteristic than the other two. Comparing to the VO$_2$, photoconductive Si can maintain insensitive to the external temperature of the surrounding environment and provides a robust switchable relative to graphene.

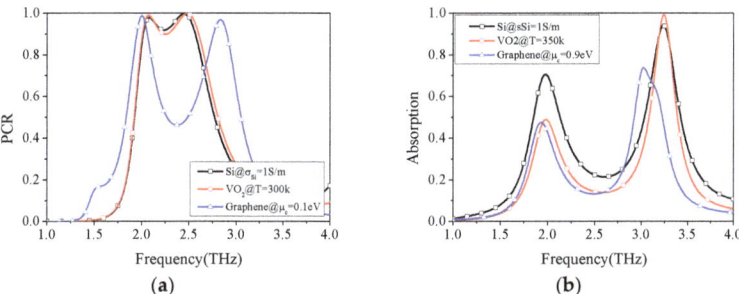

Figure 8. The PCR and absorption performances under different materials for (**a**) polarization converter and (**b**) absorber states.

4. Conclusions

A photo-excited switchable THz MTM polarization converter/absorber based on the incorporation of photoconductive Si has been designed and demonstrated in this paper. The conductivity of Si is dynamically adjusted by the external incident pump power, applied to provide a means of achieving the polarization modulation for the reflected waves. The novelty of the structure lies in its switchable performance as compared with the previous works. This design has its unique advantage of cross-polarization conversion with a relative bandwidth of 24.1% (PCR >90%) as $\sigma_{Si} = 1$ S/m without pump beam and acting as the dual-band absorber under the case of $\sigma_{Si} = 1 \times 10^5$ S/m with maximal pump optical illuminating power. The design has opened up a new field towards active switches and polarization manipulation with high performance in the THz regime. Therefore, this photo-excited switchable MTM polarization converter/absorber can be potentially applied to biological imaging, THz scanning, sensors, and so on.

Author Contributions: This study was conducted through the contributions of all authors. Conceptualization, investigation, validation, D.Y. and Y.D.; software, Y.R. and G.L. (Guochao Li); writing—original draft preparation, D.Y. and Y.D.; writing—review and editing, G.L. (Gaosheng Li) and H.M.; visualization, S.D.; supervision, Z.L. All authors have read and agreed to the published version of the manuscript.

Funding: This work was supported by the Young Creative Talents Project of Guangdong Provincial Department of Education under grant number 2020KQNCX203.

Institutional Review Board Statement: Not applicable.

Informed Consent Statement: Not applicable.

Data Availability Statement: Not applicable.

Acknowledgments: This work was supported by the Young Creative Talents Project of Guangdong Provincial Department of Education under Grant No. 2020KQNCX203.

Conflicts of Interest: The authors declare no conflict of interest.

References

1. Engheta, N. Metamaterials with high degrees of freedom: Space, time, and more. *Nanophotonics* **2020**, *10*, 639–642. [CrossRef]
2. Xiao, S.; Wang, T.; Liu, T.; Zhou, C.; Jiang, X.; Zhang, J. Active metamaterials and metadevices: A review. *J. Phys. D Appl. Phys.* **2020**, *53*, 503002. [CrossRef]
3. Deng, G.; Lv, K.; Sun, H.; Yang, J.; Yin, Z.; Li, Y.; Chi, B.; Li, X. An ultrathin, triple-band metamaterial absorber with wide-incident-angle stability for conformal applications at X and Ku frequency band. *Nanoscale Res. Lett.* **2020**, *15*, 217. [CrossRef] [PubMed]
4. Yang, D.; Yin, Y.; Zhang, Z.; Li, D.; Cao, Y. Wide-angle microwave absorption properties of multilayer metamaterial fabricated by 3D printing. *Mater. Lett.* **2020**, *281*, 128571. [CrossRef]
5. Li, H.; Ji, C.; Ren, Y.; Hu, J.; Qin, M.; Wang, L. Investigation of multiband plasmonic metamaterial perfect absorbers based on graphene ribbons by the phase-coupled method. *Carbon* **2019**, *141*, 481–487. [CrossRef]
6. Al-Badri, K.S.L. Design of perfect metamaetiral absorber for microwave applications. *Wirel. Pers. Commun.* **2021**, 1–8. [CrossRef]
7. Wang, J.; Ding, X.; Huang, X.; Wu, J.; Li, Y.; Yang, H. Metamaterials absorber for multiple frequency points within 1 GHz. *Phys. Scr.* **2020**, *95*, 065505. [CrossRef]
8. Zhou, B.C.; Wang, D.H.; Ma, J.J.; Li, B.Y.; Zhao, Y.J.; Li, K.X. An ultrathin and broadband radar absorber using metamaterials. *Waves Random Complex Media* **2021**, *31*, 911–920. [CrossRef]
9. Magisetty, R.; Raj, A.B.; Datar, S.; Shukla, A.; Kandasubramanian, B. Nanocomposite engineered carbon fabric-mat as a passive metamaterial for stealth application. *J. Alloys Compd.* **2020**, *848*, 155771. [CrossRef]
10. Li, W.; Zhang, Y.; Wu, T.; Cao, J.; Chen, Z.; Guan, J. Broadband radar cross section reduction by in-plane integration of scattering metasurfaces and magnetic absorbing materials. *Results Phys.* **2019**, *12*, 1964–1970. [CrossRef]
11. Salami, P.; Yousefi, L. Wide-band polarisation-independent metasurface-based carpet cloak. *IET Microw. Antennas Propag.* **2020**, *14*, 1983–1989. [CrossRef]
12. Yang, J.; Huang, C.; Wu, X.; Sun, B.; Luo, X. Dual-wavelength carpet cloak using ultrathin metasurface. *Adv. Opt. Mater.* **2018**, *6*, 1800073. [CrossRef]
13. Lin, M.; Xu, M.; Wan, X.; Liu, H.; Wu, Z.; Liu, J.; Deng, B.; Guan, D.; Zha, S. Single sensor to estimate DOA with programmable metasurface. *IEEE Internet Things J.* **2021**, *8*, 10187–10197. [CrossRef]

14. Cai, J.; Zhou, Y.J.; Yang, X.M. A metamaterials-loaded quarter mode SIW microfluidic sensor for microliter liquid characterization. *J. Electromagn. Waves Appl.* **2019**, *33*, 261–271. [CrossRef]
15. Ming-zhen, X.; Yang, Z.; Wei-ling, F.; Jin-chun, H. Microfludic refractive index sensor based on terahertz metamaterials. *Spectrosc. Spectr. Anal.* **2021**, *41*, 1039–1043.
16. Zhang, Y.; Tian, Y.; Zhang, Y.; Dai, L.; Liu, S.; Zhang, Y.; Zhang, H. Dual-function polarizer based on hybrid metasurfaces of vanadium dioxide and Dirac semimetals. *Opt. Commun.* **2020**, *477*, 126348. [CrossRef]
17. da Silva Paiva, J.L.; da Silva, J.P.; Campos, A.L.P.D.S.; de Andrade, H.D. Using metasurface structures as signal polarisers in microstrip antennas. *IET Microw. Antennas Propag.* **2019**, *13*, 23–27. [CrossRef]
18. Yao, Z.; Lu, M.; Zhang, C.; Wang, Y. Dynamically tunable and transmissive linear to circular polarizer based on graphene metasurfaces. *JOSA B* **2019**, *36*, 3302–3306. [CrossRef]
19. Khan, S.; Eibert, T.F. A dual-band metasheet for asymmetric microwave transmission with polarization conversion. *IEEE Access* **2019**, *7*, 98045–98052. [CrossRef]
20. Zhong, R.; Yang, L.; Liang, Z.; Wu, Z.; Wang, Y.; Ma, A.; Fang, Z.; Liu, S. Ultrawideband terahertz absorber with a graphene-loaded dielectric hemi-ellipsoid. *Opt. Express* **2020**, *28*, 28773–28781. [CrossRef]
21. Dong, Y.; Yu, D.; Li, G.; Li, G.; Ma, H. Tunable ultrathin ultrabroadband metamaterial absorber with graphene-stack-based structure at lower terahertz frequency. *Phys. E Low-Dimens. Syst. Nanostruct.* **2021**, *128*, 114608. [CrossRef]
22. Liu, H.B.; Hu, C.X.; Wang, Z.L.; Zhang, H.F.; Li, H.M. An ultra-wideband terahertz metamaterial absorber based on the fractal structure. *Plasmonics* **2020**, *16*, 263–271. [CrossRef]
23. Zhang, J.; Li, Z.; Shao, L.; Zhu, W. Dynamical absorption manipulation in a graphene-based optically transparent and flexible metasurface. *Carbon* **2021**, *176*, 374–382. [CrossRef]
24. Chu, H.; Qi, J.; Xiao, S.; Qiu, J. A thin wideband high-spatial-resolution focusing metasurface for near-field passive millimeter-wave imaging. *Appl. Phys. Lett.* **2018**, *112*, 174101. [CrossRef]
25. Zhu, W.M.; Cai, H.; Mei, T.; Bourouina, T.; Tao, J.F.; Lo, G.Q.; Kwong, D.L.; Liu, A.Q. A MEMS tunable metamaterial filter. In Proceedings of the 2010 IEEE 23rd International Conference on Micro Electro Mechanical Systems (MEMS), Hong Kong, China, 24–28 January 2010; Volume 33, pp. 196–199.
26. Ma, F.; Lin, Y.S.; Zhang, X.; Lee, C. Tunable multiband terahertz metamaterials using a reconfigurable electric split-ring resonator array. *Light Sci. Appl.* **2014**, *3*, e171. [CrossRef]
27. Kliros, G.S.; Venetis, D.E. Design of a tunable dual-band THz absorber based on metamaterial split-ring resonators coupled to grapheme. In Proceedings of the 2016 IEEE International Semiconductor Conference (CAS), Sinaia, Romania, 10–12 October 2016; pp. 41–44.
28. Liu, C.; Bai, Y.; Ma, X. Design of a tunable dual-band terahertz absorber based on graphene metamaterial. *Opt. Eng.* **2018**, *57*, 117105. [CrossRef]
29. Ahmadivand, A.; Gerislioglu, B.; Ramezani, Z. Gated graphene island-enabled tunable charge transfer plasmon terahertz metamodulator. *Nanoscale* **2019**, *11*, 8091–8095. [CrossRef] [PubMed]
30. Zhao, Y.; Huang, Q.; Cai, H.; Lin, X.; Lu, Y. A broadband and switchable VO$_2$-based perfect absorber at the THz frequency. *Opt. Commun.* **2018**, *426*, 443–449.
31. Wang, S.; Cai, C.; You, M.; Liu, F.; Wu, M.; Li, S.; Bao, H.; Kang, L.; Werner, D.H. Vanadium dioxide based broadband THz metamaterial absorbers with high tunability: Simulation study. *Opt. Express* **2019**, *27*, 19436–19447. [CrossRef]
32. Zou, H.; Cheng, Y. Design of a six-band terahertz metamaterial absorber for temperature sensing application. *Opt. Mater.* **2019**, *88*, 674–679. [CrossRef]
33. Liu, H.; Ren, G.; Gao, Y.; Zhu, B.; Wu, B.; Li, H.; Jian, S. Tunable terahertz plasmonic perfect absorber based on T-shaped InSb array. *Plasmonics* **2016**, *11*, 411–417. [CrossRef]
34. Agarwal, P.; Medwal, R.; Kumar, A.; Asada, H.; Fukuma, Y.; Rawat, R.S.; Battiato, M.; Singh, R. Ultrafast photo-thermal switching of terahertz spin currents. *Adv. Funct. Mater.* **2021**, *31*, 2010453. [CrossRef]
35. Zhao, J.; Cheng, Y.; Cheng, Z. Design of a photo-excited switchable broadband reflective linear polarization conversion metasurface for terahertz waves. *IEEE Photon. J.* **2018**, *10*, 4600210. [CrossRef]
36. Xu, Z.C.; Wu, L.; Zhang, Y.T.; Xu, D.G.; Yao, J.Q. Photoexcited blueshift and redshift switchable metamaterial absorber at terahertz frequencies. *Chin. Phys. Lett.* **2019**, *36*, 124202. [CrossRef]
37. Lang, T.; Shen, T.; Wang, G.; Shen, C. Tunable broadband all-silicon terahertz absorber based on a simple metamaterial structure. *Appl. Opt.* **2020**, *59*, 6265–6270. [CrossRef] [PubMed]
38. Bing, P.; Guo, X.; Wang, H.; Li, Z.; Yao, J. Characteristic analysis of a photoexcited tunable metamaterial absorber for terahertz waves. *J. Opt.* **2019**, *48*, 179–183. [CrossRef]
39. Song, Z.; Wang, Z.; Wei, M. Broadband tunable absorber for terahertz waves based on isotropic silicon metasurfaces. *Mater. Lett.* **2019**, *234*, 138–141. [CrossRef]
40. Zhao, X.; Wang, Y.; Schalch, J.; Duan, G.; Cremin, K.; Zhang, J.; Chen, C.; Averitt, R.D.; Zhang, X. Optically modulated ultra-broadband all silicon metamaterial terahertz absorbers. *ACS Photon.* **2019**, *6*, 830–837. [CrossRef]
41. You, X.; Upadhyay, A.; Cheng, Y.; Bhaskaran, M.; Sriram, S.; Fumeaux, C.; Withayachumnankul, W. Ultra-wideband far-infrared absorber based on anisotropically etched doped silicon. *Opt. Lett.* **2020**, *45*, 1196–1199. [CrossRef]

42. Cheng, Y.; Liu, J.; Chen, F.; Luo, H.; Li, X. Optically switchable broadband metasurface absorber based on square ring shaped photoconductive silicon for terahertz waves. *Phys. Lett. A* **2021**, *402*, 127345. [CrossRef]
43. Wu, J. A polarization insensitive dual-band tunable graphene absorber at the THz frequency. *Phys. Lett. A* **2020**, *384*, 126890. [CrossRef]
44. Cheng, Z.; Cheng, Y. A multi-functional polarization convertor based on chiral metamaterial for terahertz waves. *Opt. Commun.* **2019**, *435*, 178–182. [CrossRef]
45. Wong, H.; Wang, K.X.; Huitema, L.; Crunteanu, A. Active meta polarizer for terahertz frequencies. *Sci. Rep.* **2020**, *10*, 15382. [CrossRef]
46. Sasaki, T.; Nishie, Y.; Kambayashi, M.; Sakamoto, M.; Noda, K.; Okamoto, H.; Kawatsuki, N.; Ono, H. Active terahertz polarization converter using a liquid crystal-embedded metal mesh. *IEEE Photon. J.* **2019**, *11*, 1–7. [CrossRef]
47. Liu, X.; Liu, H.; Sun, Q.; Huang, N. Metamaterial terahertz switch based on split-ring resonator embedded with photoconductive silicon. *Appl. Opt.* **2015**, *54*, 3478–3483. [CrossRef]
48. Ding, J.; Arigong, B.; Ren, H.; Zhou, M.; Shao, J.; Lin, Y.; Zhang, H. Efficient multiband and broadband cross polarization converters based on slotted L-shaped nanoantennas. *Opt. Express* **2014**, *22*, 29143–29151. [CrossRef] [PubMed]
49. Tang, S.C.; Yu, C.H.; Chiou, Y.C.; Kuo, J.T. Extraction of electric and magnetic coupling for coupled symmetric microstrip resonator bandpass filter with tunable transmission zero. In Proceedings of the 2009 Asia Pacific Microwave Conference, Singapore, 7–10 December 2009; pp. 2064–2067.
50. Al-Behadili, A.A.; Mocanu, I.A.; Codreanu, N.; Pantazica, M. Modified split ring resonators sensor for accurate complex permittivity measurements of solid dielectrics. *Sensors* **2020**, *20*, 6855. [CrossRef] [PubMed]
51. Sheikh, S. Miniaturized-element frequency-selective surfaces based on the transparent element to a specific polarization. *IEEE Antennas Wirel. Propag. Lett.* **2016**, *15*, 1661–1664. [CrossRef]

Article

AgSn[Bi$_{1-x}$Sb$_x$]Se$_3$: Synthesis, Structural Characterization, and Electrical Behavior

Paulina Valencia-Gálvez [1], Daniela Delgado [1], María Luisa López [2], Inmaculada Álvarez-Serrano [2], Silvana Moris [3,*] and Antonio Galdámez [1,*]

[1] Departamento de Química, Facultad de Ciencias, Universidad de Chile, Las Palmeras 3425, Santiago 7800003, Chile; paulygav@uchile.cl (P.V.-G.); dani.delgado.m@gmail.com (D.D.)
[2] Departamento de Química Inorgánica, Facultad de Ciencias Químicas, Universidad Complutense, 28040 Madrid, Spain; marisal@quim.ucm.es (M.L.L.); ias@quim.ucm.es (I.Á.-S.)
[3] Centro de Investigación de Estudios Avanzados del Maule (CIEAM), Vicerrectoría de Investigación y Postgrado, Universidad Católica del Maule, Avenida San Miguel 3605, Talca 3480112, Chile
* Correspondence: smoris@ucm.cl (S.M.); agaldamez@uchile.cl (A.G.)

Citation: Valencia-Gálvez, P.;
Delgado, D.; López, M.L.;
Álvarez-Serrano, I.; Moris, S.;
Galdámez, A. AgSn[Bi$_{1-x}$Sb$_x$]Se$_3$:
Synthesis, Structural
Characterization, and Electrical
Behavior. *Crystals* 2021, 11, 864.
https://doi.org/10.3390/
cryst11080864

Academic Editors: Michael Waltl and Dmitri Donetski

Received: 13 June 2021
Accepted: 23 July 2021
Published: 26 July 2021

Publisher's Note: MDPI stays neutral with regard to jurisdictional claims in published maps and institutional affiliations.

Copyright: © 2021 by the authors. Licensee MDPI, Basel, Switzerland. This article is an open access article distributed under the terms and conditions of the Creative Commons Attribution (CC BY) license (https://creativecommons.org/licenses/by/4.0/).

Abstract: Herein, we report the synthesis, characterization, and electrical properties of lead-free AgSn$_m$[Bi$_{1-x}$Sb$_x$]Se$_{2+m}$ (m = 1, 2) selenides. Powder X-ray diffraction patterns and Rietveld refinement data revealed that these selenides consisted of phases related to NaCl-type crystal structure. The microstructures and morphologies of the selenides were investigated by backscattered scanning electron microscopy, energy-dispersive X-ray spectroscopy, and high-resolution transmission electron microscopy. The studied AgSn$_m$[Bi$_{1-x}$Sb$_x$]Se$_{2+m}$ systems exhibited typical p-type semiconductor behavior with a carrier concentration of approximately ~+10^{20} cm^{-3}. The electrical conductivity of AgSn$_m$[Bi$_{1-x}$Sb$_x$]Se$_{2+m}$ decreased from ~3.0 to ~10^{-3} S·cm^{-1} at room temperature (RT) with an increase in m from 1 to 2, and the Seebeck coefficient increased almost linearly with increasing temperature. Furthermore, the Seebeck coefficient of AgSn[Bi$_{1-x}$Sb$_x$]Se$_3$ increased from ~+36 to +50 µV·K^{-1} with increasing Sb content (x) at RT, while its average value determined for AgSn$_2$[Bi$_{1-x}$Sb$_x$]Se$_4$ was approximately ~+4.5 µV·K^{-1}.

Keywords: selenides; seebeck coefficient; lead-free thermoelectric materials

1. Introduction

The multicomponent silver chalcogenide family AgM^{1}M^{2}Q$_3$ (M^1 = Pb, Sn; M^2 = Bi, Sb, and Q = S, Se, Te) contains several thermoelectric phases [1–7]. Kanatzidis et al. extensively investigated the properties of chalcogenide compounds, which represent promising thermoelectric materials [8–10]. In particular, AgPb$_m$SbTe$_{2+m}$, Ag$_{1.33}$Pb$_{1.33}$Sb$_{1.33}$Te$_4$, and AgPb$_{18}$SbSe$_{20}$ were examined in detail. AgPb$_m$SbTe$_{2+m}$ is a class of candidates for thermoelectric materials, which is often abbreviated as LAST (lead, antimony, silver, tellurium)-m. These compounds can be envisioned as an alloy between two cubic chalcogenides, PbTe- and AgSbTe$_2$: (PbTe)$_m$–AgSbTe$_2$. Electrical transport studies have shown that the thermoelectric properties of this system can be fine-tuned via compositional parameter m. Recently, Dutta et al. and Cai et al. reported the thermoelectric properties of n-type AgPbBiSe$_3$ and p-type AgPb$_m$SbSe$_{2+m}$ families [11,12]. These phases possess high Seebeck coefficients S equal to approximately −131 µV·K^{-1} (AgPbBiSe$_3$) and +130 µV·K^{-1} (AgPb$_m$SbSe$_{2+m}$; m = 10). In addition, AgPbBiSe$_3$ exhibits S values of −131 and −204 µV·K^{-1} at 296 and 770 K, respectively, and electrical conductivity σ of 63 S·cm^{-1} at room temperature (RT) and 72 S·cm^{-1} at 818 K. As expected, these values increased with an increase in the halogen doping concentration. Experimental single-crystal X-ray diffraction (XRD) and high-resolution transmission electron microscopy (HRTEM) studies revealed that the electrical properties of these chalcogenides were related to the presence of nanoscopic inhomogeneities [4,13,14]. Electron diffraction and energy-dispersive X-ray spectroscopy

(EDS) analyses demonstrated that Ag-Sb-rich nanostructures were embedded into the PbTe lattice of (PbTe)$_m$–AgSbTe$_2$ systems, whose electronic behaviors were sensitive to their microstructures.

Owing to the strict environmental regulations related to the use of lead-based materials, novel compounds, such as lead-free p-type AgSnSbSe$_3$, n-type BiAgSeS, p-type (SnTe)$_m$–AgSbTe$_2$, p-type AgSn$_m$SbSe$_2$Te$_m$, and (SnTe)$_m$–AgBiTe$_2$ systems, have been prepared recently [15–21]. (SnTe)$_m$–AgBiTe$_2$ systems could be viewed as alloys of two cubic chalcogenides, SnTe and AgBiTe$_2$. These chalcogenides possess low thermal conductivities and large positive Seebeck coefficients. The influence of chemical substitution on the thermoelectric properties of the listed chalcogenides has been discussed in several studies. For example, the chemical substitution of Se by Te in AgSnSbSe$_3$ (e.g., AgSnSbSe$_{1.5}$Te$_{1.5}$) results in the formation of a material with an average power factor $S^2 \cdot \sigma$ of ~9.54 µW·cm^{-1}·K^{-2} (400–778 K) [15]. Herein, we report the synthesis, characterization, and electrical properties of novel lead-free AgSn$_m$[Bi$_{1-x}$Sb$_x$]Se$_{2+m}$ (m = 1 and 2) systems obtained via the isoelectronic substitution of Bi by Sb atoms. AgSn$_m$[Bi$_{1-x}$Sb$_x$]Se$_{2+m}$ compounds could be considered alloys of SnSe and Ag(Bi$_{1-x}$Sb$_x$)Se$_2$, corresponding to (SnSe)$_m$–Ag(Bi$_{1-x}$Sb$_x$)Se$_2$ systems. Tan et al. suggested that bismuth was a better neutralizer of positive holes in the SnTe structure than antimony [18]. In this study, we investigated the electrical and structural characteristics of selenides that were selected because bismuth was a more efficient electron donor than antimony. Powder X-ray diffraction patterns and Rietveld refinement results were consistent with phases related to the cubic NaCl-type lattice. The microstructures and morphologies of these systems were investigated using scanning electron microcopy (SEM) and high-resolution transmission electron microscopy (HRTEM).

2. Materials and Methods

Silver powder (99.99% purity, Sigma-Aldrich, St. Louis, MO, USA), antimony powder (99.99% purity, Sigma-Aldrich, St. Louis, MO, USA), tin powder (99.9% purity, Sigma-Aldrich, St. Louis, MO, USA), selenium powder (99.99% purity, Sigma-Aldrich, St. Louis, MO, USA), bismuth powder (99.99% purity, Merck, Kenilworth, NJ, USA), and sulfur powder (99.99% purity, Merck, Kenilworth, NJ, USA) were used in this study. All experiments were performed under a dry and oxygen-free argon atmosphere. Selenide phases with the nominal compositions were prepared via the solid-state reaction of Ag, Sn, Bi, Sb, and Se powders mixed in the stoichiometric proportions inside evacuated quartz ampoules. The reaction mixture was gradually heated to 1223 K at a rate of +50 K/h, maintained at this temperature for ~12 h, and slowly cooled to RT at 8 K/h. Chemical compositions of the samples were determined by scanning electron microscopy (SEM, JEOL 5400 system, Tokyo, Japan) and energy-dispersive X-ray spectroscopy (EDS, Oxford LinK ISIS microanalyzer, Oxford Instruments, Abingdon, UK). XRD patterns were obtained at RT using a Bruker D8 advanced powder diffractometer (Bruker, Billerica, MA, USA) with CuKα radiation over the 2θ range of 5–80° at a step size of 0.01°. In addition, XRD patterns were also collected at temperatures varying from RT to 77 K using a PANalytical X'Pert PROMPD diffractometer (CuKα1 = 1.544426 Å, CuKα2 = 1.54098 Å) over the 2θ range of 5–80° at a step size of 0.0167° (Malvern Panalytical, Boulder, CO, USA). The diffractometer was equipped with a heating chamber (Anton Paar HTK1200, PANanalytical BV, Boulder, CO, USA). The collected data were analyzed by two Rietveld refinement software programs: Fullprof and MAUD [22–24]. A standard LaB$_6$ sample was used to measure instrumental profiles. SEM–EDS and powder XRD analyses revealed that the AgSn$_m$[Bi$_{1-x}$Sb$_x$]Se$_{2+m}$ (m = 1 and 2) samples have been successfully prepared. On the other hand, reaction products with the nominal stoichiometry of AgSn$_m$[Bi$_{1-x}$Sb$_x$]Se$_{m+2}$ (m = 4, 8, 10, and 12) included 10–30% SnSe impurities. For example, the reaction products with the nominal compositions AgSn$_8$[Bi$_{0.8}$Sb$_{0.2}$]Se$_{10}$ and AgSn$_4$[Bi$_{0.2}$Sb$_{0.8}$]Se$_6$ contained AgSnMSe$_3$, SnSe, and some unidentified impurities. Differential thermal analysis (DTA) and thermogravimetric analysis (TGA) of the samples were performed using a Rheometric Scientific STA 1500H/625 thermal analysis system (Rheometric Scientific, Inc., Piscataway, NJ, USA).

To record DTA/TGA curves, the samples were heated from RT to 1273 K in an argon atmosphere or to 673 K in air at a rate of 10 K/min. A high-temperature melting method was applied to obtain suitable ingot samples for Seebeck measurements. The obtained reaction samples were crushed into powders and placed into a quartz ampoule, which was evacuated and flame-sealed under an argon atmosphere. This tube was then placed into a furnace at 1073 K for ~2 h and then slowly cooled to RT at 8 K/h. Subsequently, these ingots were cut and polished for measurements of their transport properties. Although various experimental conditions (temperature, cooling rate, melting, and cooling) were utilized, the ingots of some chemical compositions of $AgSn_2[Bi_{1-x}Sb_x]Se_4$ and $AgSn[Bi_{1-x}Sb_x]Se_3$ were too brittle and broken. In the case of the $AgSn_2[Bi_{1-x}Sb_x]Se_4$ composition, a visible change in the pellet appearance (the presence of small holes on the pellet surface) was observed in some samples during thermal cycling. Electrical conductivities of the samples were measured by a pellet method, in which pellets were uniaxially pressed at 5×10^8 Pa to form cylindrical specimens with diameters of ~7 mm and thicknesses of ~1–2 mm. The pelletized specimens were sintered at 673, 923, and 993 K for 12 and 24 h in an argon atmosphere (Ar 12 or 24 h) and in air at 673 K for 12 h (air 12 h). The low-temperature (20–300 K) Hall effect and Seebeck coefficient measurements were performed using a physical property measurement system (PPMS). High-temperature electrical conductivities were measured using an ECOPIA HMS 2000 system (ECOPIA, Anyang-city, Gyeonggi-do, South Korea) and Keithley 6220 current source/Keithley 2182A nanovoltmeter (Tektronix, Inc., Beaverton, OR, USA) with a four-probe contact geometry. High-temperature Seebeck coefficients were determined by a laboratory-made system. The densities of the ingot and pellet samples were computed from their dimensions and masses; the obtained magnitudes were equal to ~90–92% of the theoretical values.

3. Results and Discussion

3.1. Structures and Compositions of Selenide Samples

The backscattered electron images and EDS analyses of the powder samples, ingots, and pellets for electrical measurements revealed that the chemical compositions of all selenide samples were uniform throughout the scanned region (Figure 1). No secondary intergranular phases were observed in the $AgSn[Bi_{1-x}Sb_x]Se_3$ samples (x = 0, 0.2, 0.3, 0.8) within the detection limits of this technique. Figure 2 presents the PXRD patterns for the $AgSn_m[Bi_{1-x}Sb_x]Se_{2+m}$ samples with m = 1 and 2. Their sharpness indicates high crystallinity, and the observed interlayer spacing was in good agreement with the calculated interplanar spacing d. The SEM–EDS and XRD analyses suggest that the prepared materials could be regarded as a continuous solid solution between SnSe and $AgBiSe_2$, similarly to the $SnTe-NaSbTe_2$ and $SnTe-NaBiTe_2$ systems [25]. In the case of $AgSn_2[Bi_{1-x}Sb_x]Se_4$ samples (m = 2), the PXRD patterns show an extremely small amount of SnSe impurities.

A structural analysis of the $AgPb_mSbTe_{2+m}$ phases carried out by Quarez et al. by HRTEM analyses and single-crystal XRD demonstrated that the structure of these phases could be refined in the space group $Pm\bar{3}m$, as well as in a lower symmetric space group such as $P4/mmm$ [13]. It is extremely complicated to choose the adequate symmetry since all refinements were acceptable due to the coherent X-ray scattering amplitudes, which are extremely similar, and the super-structure reflections in the powder pattern are especially hard to see.

The XRD patterns obtained for these phases were refined in the Pm-$3m$ and $P4/mmm$ space groups. In both space groups, the two sites at the (0,0,1/2) and (1/2,1/2,0) positions were fully occupied by Ag, Sb, Sn, or Bi atoms. The RT Rietveld refinement profiles of $AgSnBiSe_3$ based on the cubic and tetragonal models are shown in Figure 3. The following unit cell parameters were obtained: cubic unit cell a_c = 5.86180(6) Å, tetragonal ($a_t = b_t \approx a_c\sqrt{2}/2$, $c_t \approx a_c$) unit cell a_t = 4.14365 (16), and c_t = 5.8631(4) Å. The tetragonal model contained two cation sites (1b and 1c Wyckoff sites), which were fully occupied by the randomly distributed Ag, Sn, and Bi cations. Selenium atoms adopted close cubic packing at the 1a and 1d Wyckoff sites. Both structural models were consistent with the experimental data as well as with the obtained R factors; therefore, we were unable to identify the best structural model for this material (Tables S1 and S2). Moreover, AgBiSe$_2$ and Ag(Bi,Sb)Se$_2$ phases show interesting structural transitions with both temperature and composition. Thus, at room temperature the Sb-free samples have a hexagonal symmetry within the space group $P\bar{m}31$. As the Sb content increases, these phases show a rhombohedral symmetry ($R\bar{3}m$) [26].

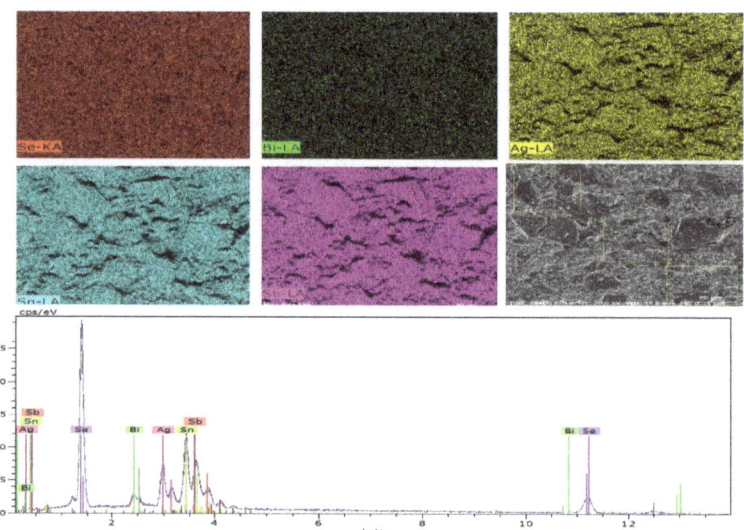

Figure 1. SEM analysis: representative EDS mapping images (powder sample) and backscattered electron image of AgSn[Bi$_{0.8}$Sb$_{0.2}$]Se$_3$ (20 kV, 670×).

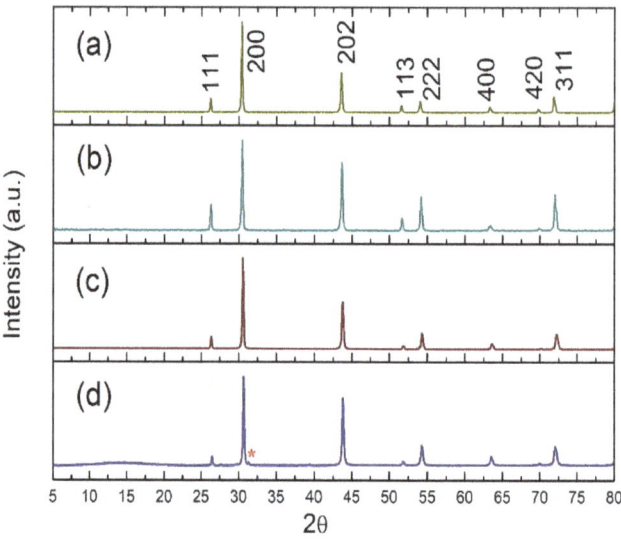

Figure 2. Representative powder XRD patterns at room temperature of: (**a**) AgSnBiSe$_3$ showing the corresponding *hkl* miller indices, (**b**) AgSn[Bi$_{0.8}$Sb$_{0.2}$]Se$_3$, (**c**) AgSn[Bi$_{0.2}$Sb$_{0.8}$]Se$_3$, and (**d**) AgSn$_2$[Bi$_{0.2}$Sb$_{0.8}$]Se$_4$ (the red asterisk indicates a secondary reflection indexed to the SnSe *Pnma* space group).

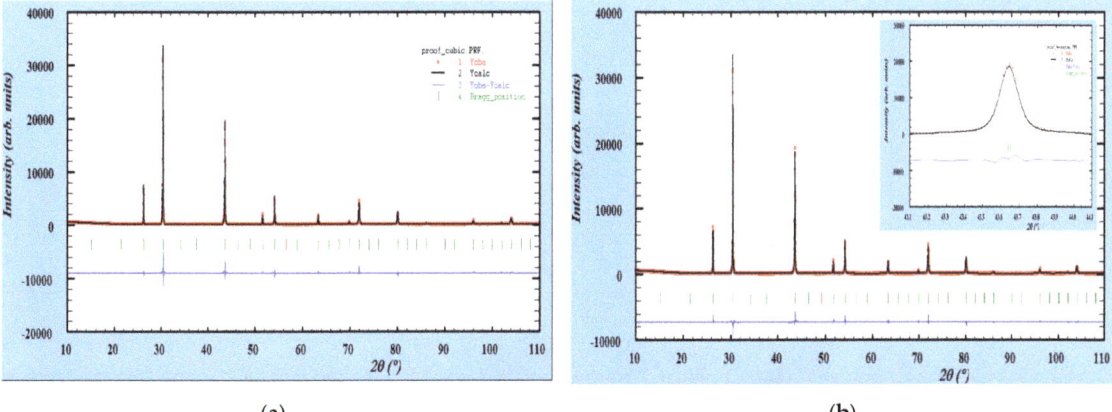

Figure 3. Rietveld refinement of the XRD data obtained for AgSnBiSe$_3$ (including the profile fit, profile difference, and profile residuals), which was performed with the Fullprof software. (**a**) Cubic *Pm-3m* and (**b**) tetragonal *P4/mmm* space groups at RT (the insert shows a selected area). Bragg's R-factors of the cubic space group were $Rb = 3.32$ and $Rwp = 4.44$. Bragg's R-factors of the tetragonal space group were $Rb = 5.72$ and $Rwp = 4.88$.

XRD patterns of AgSn[Bi$_{1-x}$Sb$_x$]Se$_3$ and AgSn$_2$[Bi$_{1-x}$Sb$_x$]Se$_4$ were refined using the cubic and tetragonal models (Table S3 and Figure S1—see in Supplementary Materials). As expected, the cell volume increased slightly after replacing Bi atoms (radius: 1.63 Å) by Sb atoms (radius: 1.53 Å), and the resulting solid solutions did not obey Vegard's law.

In order to analyze the microstructural features of the samples, a study by HRTEM and electron diffraction was also carried out. Small crystals with octahedral AgSnBiSe$_3$ shapes were observed. The mean atomic compositions (determined from the EDS data) of the studied samples were similar to the expected compositions. Figure 4 displays the electron diffraction (ED) patterns obtained along [111]c zone axis for the different selected regions of the selenide samples. It shows that different contrast zones are predominant in the analyzed crystals, as shown in Figure 4.

The Miller indices are calculated considering a cubic symmetry in both models and they are indicated in the image. However, spots of weak intensity that are prohibited in the space group $Fm\overline{3}m$ are clearly observed. It is noteworthy that these weak spots indicate the appearance of different symmetries. The new Miller indices can be obtained from the Equations (1) and (2) where the unit cell transformation from cubic to tetragonal or hexagonal were considered.

Equation (1): cubic to tetragonal.

$$\begin{aligned} h_t &= \tfrac{1}{2}(h_c - k_c) \\ K_t &= \tfrac{1}{2}(h_c + k_c) \\ l_t &= l_c \end{aligned} \qquad (1)$$

Equation (2): cubic to hexagonal

$$\begin{aligned} h_h &= \tfrac{1}{2}(-k_c + l_c) \\ K_h &= \tfrac{1}{2}(h_c - l_c) \\ l_h &= h_c + k_c + l_c \end{aligned} \qquad (2)$$

Thus, in the micrograph of Figure 4a, these weak spots would not be compatible with a P-type cubic lattice, nor with a tetragonal symmetry, as shown in the diagrams below the ED images. However, they would be compatible with a hexagonal symmetry. On the contrary, only a $Pm\overline{3}m$ cubic space group seems to be consistent with the superstructure spots (Figure 4b). These results indicate that the AgSnBiSe$_3$ phase exhibits nanoscopic inhomogeneities, in good agreement with those reported for similar phases [10,13,27]. In addition, nanoregions with different orientations or symmetries were detected (Figure 4c). This is consistent with the results reported previously [13].

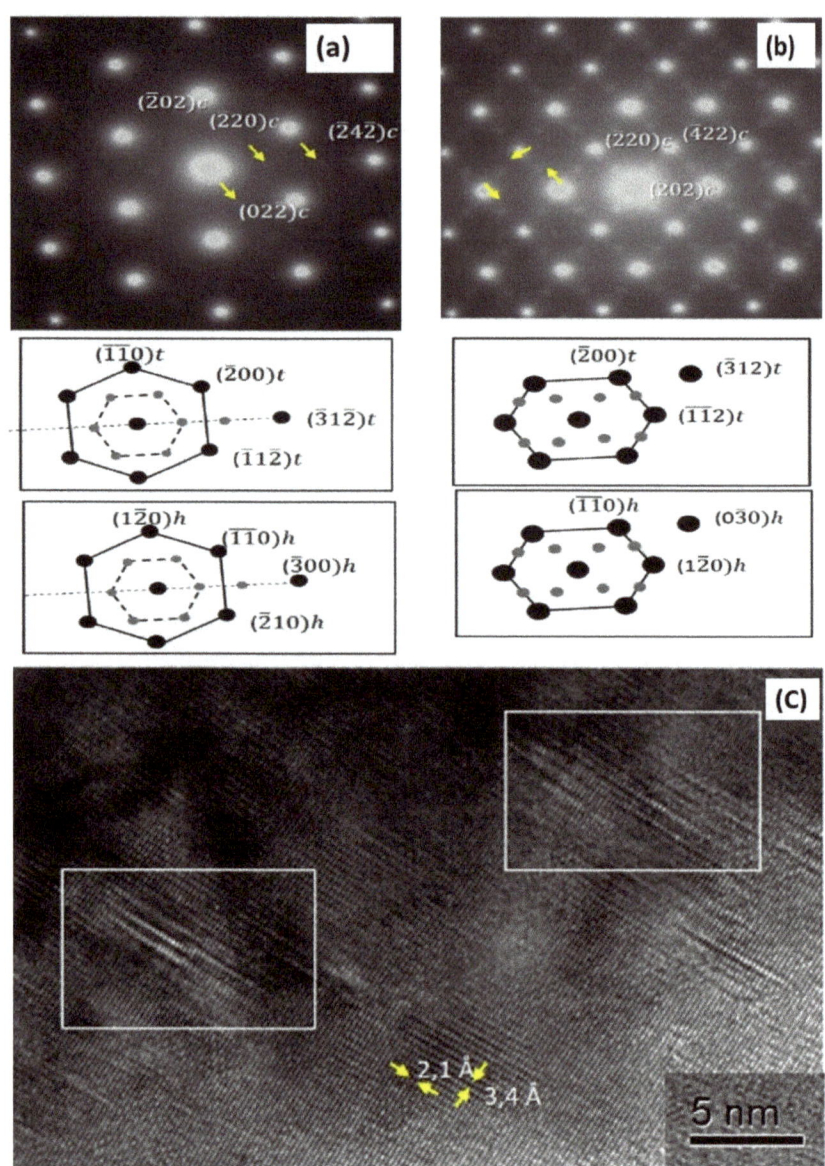

Figure 4. High-resolution transmission electron microscopy (HRTEM) study for AgSnBiSe$_3$, showing (a,b) ED patterns along [111]c zone axis and schemes for different symmetries (h and t refer to hexagonal and tetragonal, respectively), and (c) representative HRTEM image showing disorder regions apparent at the nanoscale.

3.2. Electrical Properties and Seebeck Coefficients

The melting point of the prepared selenide samples was ~1000 K, and the pellets were sintered at 673, 923, and 993 K (see the experimental details). In addition, TGA was conducted in air from RT to 673 K to produce stable thermal and gravimetric signals. The obtained TGA results were consistent with the SEM–EDS and powder XRD data (Figures S2 and S3). The AgSn[Bi$_{1-x}$Sb$_x$]Se$_3$ pellets exhibited a microstructure with particle sizes of ~2–14 µm and isolated regions with a grain size of ~20 µm. The XRD patterns of the selenide samples recorded at various temperatures between RT and 77 K exhibit high structural and thermal stabilities (Figure S4). Their XRD and backscattered SEM–EDX analyses were performed after electrical measurements.

Figure 5a shows the temperature dependence of the electrical conductivity σ of the AgSnBiSe$_3$ sample. For pristine AgSnBiSe$_3$, electrical conductivity increases with increasing temperature, indicating the phase exhibits semiconductor behavior. The σ was thermally activated at low temperatures. In addition, the σ of the samples sintered at 923 K in Ar for 48 h was ~2.1 S·cm^{-1} at 240 K. This value was approximately two times higher than the σ of the samples sintered at 673 K in Ar or air (~1.1 S·cm^{-1}) at 240 K. The AgSnBiSe$_3$ sample with a cylindrical shape (ϕ~6.00 mm and s~1.0 mm) cut from the ingot of pristine material, exhibited an electrical conductivity of ~4.0 S·cm^{-1} at RT. The σ magnitudes obtained at 80 K and 140 K were equal to 0.24 and 0.75 S·cm^{-1}, respectively. The electrical conductivity σ of AgSn$_m$[Bi$_{1-x}$Sb$_x$]Se$_{2+m}$ decreased from ~3.0 to 10^{-3} S·cm^{-1} at RT with an increase in m from 1 to 2 (Table 1). These values are comparable to those of AgSn$_m$SbSe$_{m+2}$ (~10^{-3} – 93 S·cm^{-1} for m = 1–12) [15].

Table 1. Electrical conductivity (σ) and Seebeck coefficient of AgSn[Bi$_{1-x}$Sb$_x$]Se$_3$ and AgSn$_2$[Bi$_{1-x}$Sb$_x$]Se$_4$.

	σ at RT (S·cm^{-1}) [£]	Seebeck Coefficient at RT (µV·K^{-1}) [§]	Seebeck Coefficient at 450 K (µV·K^{-1}) [§] *
AgSnBiSe$_3$	2.46	+7.39	+89.6
AgSnBi$_{0.8}$Sb$_{0.2}$Se$_3$	3.22	+36.2	-
AgSnBi$_{0.2}$Sb$_{0.8}$Se$_3$	2.97	+49.8	-
AgSn$_2$Bi$_{0.8}$Sb$_{0.2}$Se$_4$	1.22 × 10^{-3}	+3.93	+92.0
AgSn$_2$Bi$_{0.2}$Sb$_{0.8}$Se$_4$	2.35 × 10^{-3}	+4.91	+114.1

[£] pelletized specimens and [§] ingot samples. * Because of the brittleness and chemical unstable nature of some samples, the Seebeck coefficients at 450 K were not reported in this table (see details in materials and methods section). The densifications magnitudes were equal to ~90–92%.

The low-temperature $\sigma(T)$ of AgSnBiSe$_3$ followed a standard variable range hopping (VRH) model described by the following equation:

$$\sigma = \sigma_0 \exp\left[-\left(\frac{T_0}{T}\right)^{1/4}\right] \quad (3)$$

where σ_0 is the residual conductivity, and T_o is the characteristic temperature. To validate the VRH model, we fitted the experimental data with Equation (3). The obtained results presented in the inset of Figure 5a indicate that the transport mechanism regime involved the VRH process. The plot of ln σ versus $T^{-1/4}$ changed its the slope at around ~160 K. The occurrence of the hopping process depends on the energy difference and relationship between the Fermi level and the mobility edge. This behavior was previously observed at low temperatures for other selenides, such as Pb-doped Cu$_2$SnSe$_3$ and Cu$_y$Fe$_4$Sn$_{12}$X$_{32}$ spinels [28,29].

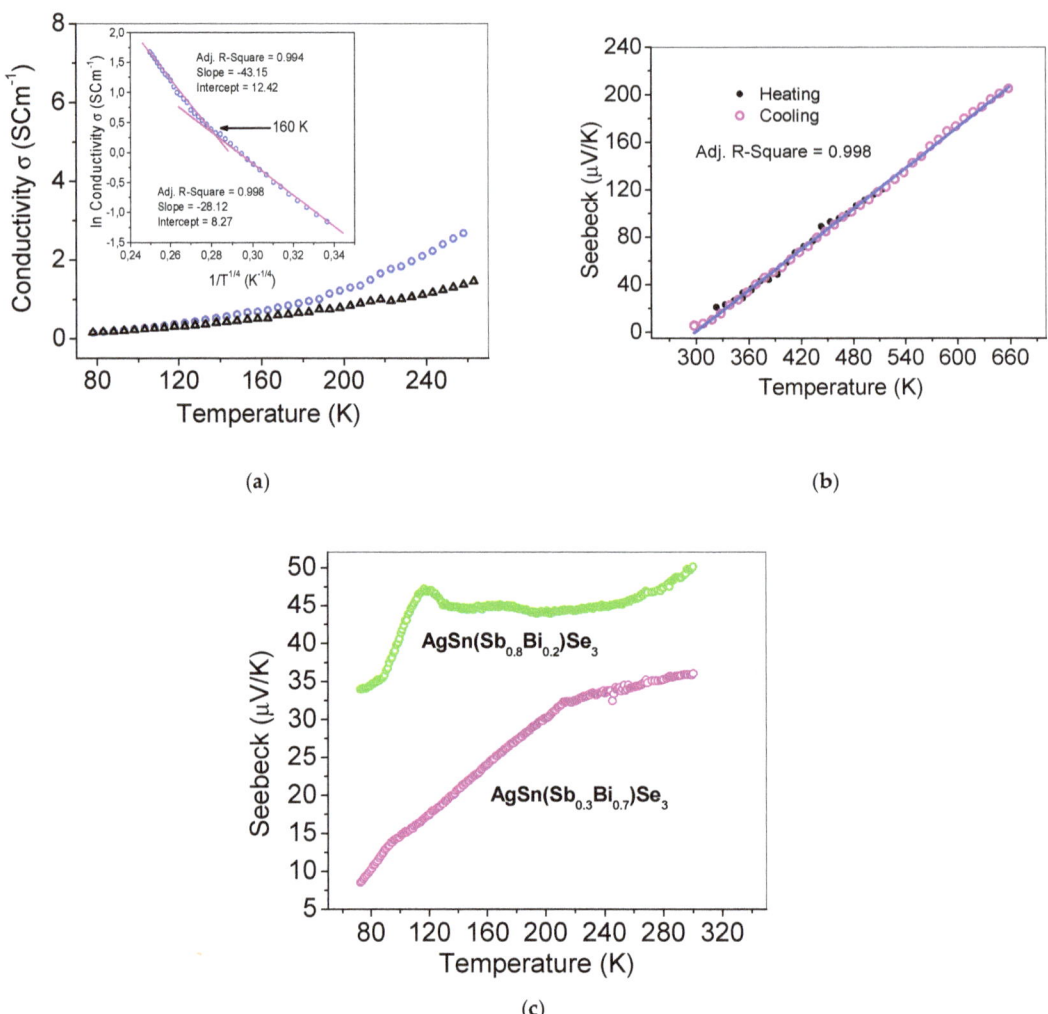

Figure 5. Temperature dependences of the electrical conductivity and Seebeck coefficient determined for AgSn[Bi$_{1-x}$Sb$_x$]Se$_3$. (a) Electrical conductivity–temperature plots of AgSnBiSe$_3$ sintered at 673 K in Ar 12 h (open triangles) and AgSnBiSe$_3$ sintered at 923 K in Ar for 48 h (open circles). The inset shows the ln σ vs. T$^{-1/4}$ curve, and the solid lines denote the linear fit of the experimental data. (b) Plots of Seebeck coefficient vs. temperature (heating–cooling cycle measurements) obtained for AgSnBiSe$_3$. (c) Plots of Seebeck coefficient vs. temperature obtained for AgSn[Bi$_{1-x}$Sb$_x$]Se$_3$ ($x = 0.8$ and 0.3) in the low-temperature range.

The positive Seebeck coefficient values of the AgSn$_m$[Bi$_{1-x}$Sb$_x$]Se$_{2+m}$ systems ($m = 1$ and 2) obtained via the isoelectronic substitution of a fraction of Bi atoms with Sb indicates that holes are the dominant conduction carriers (Table 1). The temperature dependences of the Seebeck coefficient of AgSn[Bi$_{1-x}$Sb$_x$]Se$_3$ in the high-temperature and low-temperature ranges are shown in Figure 5b,c, respectively. The Seebeck coefficient of AgSn[Bi$_{1-x}$Sb$_x$]Se$_3$ ($m = 1$) selenides increased from ~+36 µV·K^{-1} ($x = 0.2$) to +50 µV·K^{-1} ($x = 0.8$) with increasing Sb content at RT. This value is approximately four times lower than that of AgSnSbSe$_3$ (+179 µV·K^{-1}) and is comparable with the Seebeck coefficient of SnTe at 300 K (+40 µV K^{-1}). Consequently, the power factor ($S^2 \cdot \sigma$) of AgSnBiSe$_3$ was ~0.1 µW·cm^{-1}·K^{-2}. Power factors of 5–12 µW·cm^{-1}·K^{-2} were obtained for AgSn$_m$SbTe$_{m+2}$ systems at 300 K [21]. The S values of AgSn$_2$[Bi$_{1-x}$Sb$_x$]Se$_4$ selenides increased from ~+3.9 (x = 0.2) to ~+4.9 µV·K^{-1} (x = 0.8) with an increase in the Sb content at RT (Table 1). The average Seebeck coefficient of the AgSn$_2$[Bi$_{1-x}$Sb$_x$]Se$_4$ samples was ~+4.5 µV·K^{-1}. This value is approximately ten times lower than that determined for AgSn[Bi$_{0.2}$Sb$_{0.8}$]Se$_3$ (~+50 µV·K^{-1}). Kanatzidis et al. reported that the Seebeck coefficients of p-type AgSn$_m$SbSe$_{2+m}$ systems decreased from ~+179 to 51 µV·K^{-1} with an increase in m from 1 to 12 at RT [15]. The AgSn$_2$[Bi$_{1-x}$Sb$_x$]Se$_4$ selenide samples exhibited S values of ~+92 µV·K^{-1} (x = 0.2) and +114 µV·K^{-1} (x = 0.8) at 450 K. The temperature dependence of S obtained for AgSn$_2$[Bi$_{1-x}$Sb$_x$]Se$_4$ indicate degenerate semiconductor properties and is consistent with the behavior previously reported for AgSn$_{10}$SbSe$_2$Te$_{10}$ [17].

The temperature dependence of the Seebeck coefficient of degenerate semiconductors is expressed by the formula

$$S = \left[\frac{8\pi^{2/3}k_B^2(r+3/2)}{3^{5/3}eh^2}\right]\left(\frac{m^*}{n^{2/3}}\right)T \tag{4}$$

where S is the Seebeck coefficient, m^* is the effective mass, k_B is Boltzmann's constant, e is the charge of an electron, h is Planck's constant, and n is the carrier concentration [17]. In this study, the Seebeck coefficient was fitted in the temperature range of 300 K $\leq$ T $\leq$ 700 K to determine the density-of-states effective mass (m^*). Figure 5b shows the plot of Seebeck coefficient vs. temperature constructed for AgSnBiSe$_3$ in the temperature range from RT to 670 K. The results of Hall measurements revealed that the carrier concentration in the selenide samples was approximately ~+10^{20} cm^{-3} at high temperatures. The results of Seebeck measurements conducted for AgSnBiSe$_3$ implied a large m^*~7·m_0 with an acoustic phonon scattering mechanism ($r = -1/2$) and m^*~3·m_0 with an ionized impurity scattering mechanism ($r = 3/2$). Large density-of-states effective mass, m^*~4.7−1.6·m_0 and m^*~6·m_0, were obtained for AgSn$_{10}$SbSe$_2$Te$_{10}$ and Ag$_{0.85}$SnSb$_{1.15}$Te$_3$, respectively [17,30] In addition, large values of m^* are usually obtained by the single parabolic band (SPB) model in degenerate samples. However, measurements of carrier concentration and mobilities at low temperature would be required to support this model.

4. Conclusions

SEM–EDS and powder XRD analyses revealed that the AgSn$_m$[Bi$_{1-x}$Sb$_x$]Se$_{2+m}$ (m = 1 and 2) samples have been successfully prepared at high temperatures via solid-state reactions. The HRTEM and XRD analysis of AgSnBiSe$_3$ indicated the presence of nanoregions with different symmetries, in good agreement with that reported for similar phases. The electrical conductivity of p-type AgSn$_m$[Bi$_{1-x}$Sb$_x$]Se$_{2+m}$ semiconductor decreased with an increase in m from 1 to 2. The linear plot of ln σ vs. $T^{-1/4}$ obtained for AgSnBiSe$_3$ indicated that the transport mechanism in the low-temperature regime included the VRH process. The positive Seebeck coefficients of the suggested that holes were the dominant conduction carriers. The Seebeck coefficient of AgSn[Bi$_{1-x}$Sb$_x$]Se$_3$ (m = 1) selenides increased from ~+36 (x = 0.2) to ~+50 µV·K^{-1} (x = 0.8) at RT with an increase in x.

Supplementary Materials: The following materials are available online at https://www.mdpi.com/article/10.3390/cryst11080864/s1. Table S1 and S2: Final atomic parameters determined through Rietveld refinement using the Fullprof program; Figure S1: Powder XRD data obtained for the AgSn$_m$[Bi$_{1-x}$Sb$_x$]Se$_{m+2}$ samples from the corresponding Rietveld refinement data; Table S3: Cell parameters of AgSn[Bi$_{1-x}$Sb$_x$]Se$_3$ and AgSn$_2$[Bi$_{1-x}$Sb$_x$]Se$_4$; Figure S2: SEM–BS images and ED spectrum of AgSnBiSe$_3$; Figure S3: XRD patterns of the sintered samples; Figure S4: Representative XRD patterns of the samples obtained upon heating/cooling from RT (bottom) to 77 K.

Author Contributions: Conceptualization, A.G. and S.M.; methodology and experiments, M.L.L., I.Á.-S., A.G., and D.D.; writing—original draft preparation, S.M., P.V.-G., M.L.L., I.Á.-S., and A.G.; electrical measurements, M.L.L. and P.V.-G. All authors have read and agreed to the published version of the manuscript.

Funding: This research received no external funding.

Acknowledgments: This work was supported by Fondecyt No. 1190856. The authors also acknowledge the CAI center of UCM (HRTEM). We would like to thank Daniela Ruiz and Daniela Herrera for their invaluable contributions to the synthesis of the samples and discussion of the experimental results.

Conflicts of Interest: The authors declare no conflict of interest.

References

1. Zeier, W.G.; Zevalkink, A.; Gibbs, Z.; Hautier, G.; Kanatzidis, M.; Snyder, J. Thinking Like a Chemist: Intuition in Thermoelectric Materials. *Angew. Chem. Int. Ed.* **2016**, *55*, 6826–6841. [CrossRef]
2. Zhang, X.; Zhao, L.-D. Thermoelectric materials: Energy conversion between heat and electricity. *J. Mater.* **2015**, *1*, 92–105. [CrossRef]
3. Minnich, A.J.; Dresselhaus, M.S.; Ren, Z.F.; Chen, G. Bulk nanostructured thermoelectric materials: Current research and future prospects. *Energy Environ. Sci.* **2009**, *2*, 466–479. [CrossRef]
4. Vineis, C.H.; Shakouri, A.; Majumdar, A.; Kanatzidis, M. Nanostructured Thermoelectrics: Big Efficiency Gains from Small Features. *Adv. Mater.* **2010**, *22*, 3970–3980. [CrossRef] [PubMed]
5. Hsu, K.F.; Loo, S.; Guo, F.; Chen, W.; Dyck, J.S.; Uher, C.; Hogan, T.; Polychroniadis, E.K.; Kanatzidis, M.G. Cubic $AgPb_mSbTe_{2+m}$: Bulk thermoelectric materials with high figure of merit. *Science* **2004**, *303*, 818–821. [CrossRef] [PubMed]
6. Pan, L.; Mitra, S.; Zhao, L.-D.; Shen, Y.; Wang, Y.; Felser, C.; Berardan, D. The Role of Ionized Impurity Scattering on the Thermoelectric Performances of Rock Salt $AgPb_mSnSe_{2+m}$. *Adv. Funct. Mater.* **2016**, *26*, 5149–5157. [CrossRef]
7. Xiao, Y.; Chang, C.H.; Zhang, X.; Pei, Y.; Li, F.; Yuan, B.; Gong, S.H.; Zhao, L.-D. Thermoelectric transport properties of $Ag_mPb_{100}Bi_mSe_{100+2m}$ system. *J. Mater. Sci. Mater. Electron.* **2016**, *27*, 2712–2717. [CrossRef]
8. Sootsman, J.; Chung, D.Y.; Kanatzidis, M. New and old concepts in thermoelectric materials. *Angew. Chem. Int. Ed.* **2009**, *48*, 8616–8639. [CrossRef] [PubMed]
9. Kanatzidis, M. Nanostructured thermoelectrics: The new paradigm? *Chem. Mater.* **2010**, *22*, 648–659. [CrossRef]
10. Slade, T.J.; Grovogui, J.; Kuo, J.; Anand, S.; Bailey, T.; Wood, M.; Uher, C.; Snyder, J.; Dravid, V.; Kanatzidis, M. Understanding the thermally activated charge transport in $NaPb_mSbQ_{m+2}$ (Q= S, Se, Te) thermoelectrics: Weak dielectric screening leads to grain boundary dominated charge carrier scattering. *Energy Environ. Sci.* **2020**, *13*, 1509–1518. [CrossRef]
11. Dutta, M.; Pal, K.; Waghmare, U.V.; Biswas, K. Bonding heterogeneity and lone pair induced anharmonicity resulted in ultralow thermal conductivity and promising thermoelectric properties in n-type $AgPbBiSe_3$. *Chem. Sci.* **2019**, *10*, 4905–4913. [CrossRef]
12. Cai, K.F.; He, X.R.; Avdeev, M.; Yu, D.H.; Cui, J.L.; Li, H. Preparation and thermoelectric properties of $AgPb_mSbSe_{m+2}$ materials. *J. Solid State Chem.* **2008**, *181*, 1434–1438. [CrossRef]
13. Quarez, E.; Hsu, K.-F.; Pcionek, R.; Frangis, N.; Polychroniadis, E.K.; Kanatzidis, M. Nanostructuring, compositional fluctuations, and atomic ordering in the thermoelectric materials $AgPb_mSbTe_{2+m}$. The myth of solid solutions. *J. Am. Chem. Soc.* **2005**, *127*, 9177–9190. [CrossRef] [PubMed]
14. Lioutas, C.H.; Frangis, N.; Todorov, I.; Chung, D.; Kanatzidis, M. Understanding nanostructures in thermoelectric materials: An electron microscopy study of $AgPb_{18}SbSe_{20}$ crystals. *Chem. Mater.* **2010**, *22*, 5630–5635. [CrossRef]
15. Luo, Y.; Hao, S.; Cai, S.; Slade, T.J.; Luo, Z.Z.; Dravid, V.P.; Yan, Q.; Kanatzidis, M.G. High Thermoelectric Performance in the New Cubic Semiconductor $AgSnSbSe_3$ by High-Entropy Engineering. *J. Am. Chem. Soc.* **2020**, *142*, 15187–15198. [CrossRef]
16. Pei, Y.L.; Wu, H.; Sui, J.; Li, J.; Berardan, D.; Barreteau, C.; Pan, L.; Dragoe, N.; Liu, W.S.; He, J.; et al. High thermoelectric performance in n-type BiAgSeS due to intrinsically low thermal conductivity. *Energy Environ. Sci.* **2013**, *6*, 1750–1755. [CrossRef]
17. Figueroa-Millon, S.; Álvarez-Serrano, I.; Bérardan, D.; Galdámez, A. Synthesis and transport properties of p-type lead-free $AgSn_mSbSe_2Te_m$ thermoelectric systems. *Mater. Chem. Phys.* **2018**, *211*, 321–328. [CrossRef]
18. Tan, G.; Shi, F.; Sun, H.; Zhao, L.D.; Uher, C.; Dravid, V.P.; Kanatzidis, M.G. $SnTe$–$AgBiTe_2$ as an efficient thermoelectric material with low thermal conductivity. *J. Mater. Chem. A* **2014**, *2*, 20849–20854. [CrossRef]
19. Falkenbach, O.; Schmitz, A.; Dankwort, T.; Koch, G.; Kienle, L.; Mueller, E.; Schlecht, S. Tin Telluride-Based Nanocomposites of the Type $AgSn_mBiTe_{2+m}$ (BTST-m) as Effective Lead-Free Thermoelectric Materials. *Chem. Mater.* **2015**, *27*, 7296–7305. [CrossRef]
20. Xing, Z.B.; Li, Z.Y.; Tan, Q.; Wei, T.R.; Wu, C.F.; Li, J.F. Composition optimization of p-type $AgSn_mSbTe_{m+2}$ thermoelectric materials synthesized by mechanical alloying and spark plasma sintering. *J. Alloys Compd.* **2014**, *615*, 451–455. [CrossRef]
21. Han, M.K.; Androulakis, J.; Kim, S.J.; Kanatzidis, M.G. Lead-Free Thermoelectrics: High Figure of Merit in p-type $AgSn_mSbTe_{m+2}$. *Adv. Energy Mater.* **2012**, *2*, 157–161. [CrossRef]
22. Rodriguez-Carvajal, J. Recent Advances in Magnetic Structure Determination by Neutron Powder Diffraction. *Phys. B* **1993**, *192*, 55–69. [CrossRef]
23. Rietveld, H.M. A Profile Refinement Method for Nuclear and Magnetic Structures. *J. Appl. Cryst.* **1969**, *2*, 65–71. [CrossRef]
24. Lutterotti, L. Total Pattern Fitting for the Combined Size–Strain–Stress–Texture Determination in Thin Film Diffraction. *Nucl. Inst. Methods Phys. Res. B* **2010**, *268*, 334–340. [CrossRef]

25. Slade, T.J.; Pal, K.; Grovogui, J.A.; Bailey, T.P.; Male, J.; Khoury, J.F.; Zhou, X.; Chung, D.Y.; Snyder, J.; Uher, C.; et al. Contrasting SnTe–NaSbTe$_2$ and SnTe–NaBiTe$_2$ Thermoelectric Alloys: High Performance Facilitated by Increased Cation Vacancies and Lattice Softening. *J. Am. Chem. Soc.* **2020**, *142*, 12524–12535. [CrossRef] [PubMed]
26. Sudo, K.; Goto, Y.; Sogabe, R.; Hoshi, K.; Miura, A.; Moriyoshi, C.; Kuroiwa, Y.; Mizuguchi, Y. Doping-Induced Polymorph and Carrier Polarity Changes in Thermoelectric Ag(Bi,Sb)Se$_2$ Solid Solution. *Inorg. Chem.* **2019**, *58*, 7628–7633. [CrossRef]
27. Bilc, D.; Mahanti, S.D.; Quarez, E.; Hsu, K.F.; Pcionek, R.; Kanatzidis, M.G. Resonant States in the Electronic Structure of the High Performance Thermoelectrics AgPb$_m$SbTe$_{2+m}$: The Role of Ag-Sb Microstructures. *Phys. Rev. Lett.* **2004**, *93*, 146403. [CrossRef] [PubMed]
28. Prasad, S.; Rao, A.; Gahtori, B.; Bathula, S.; Dhar, A.; Chang, C.C.; Kuo, Y.K. Low-temperature thermoelectric properties of Pb doped Cu$_2$SnSe$_3$. *Phys. B Condens. Matter* **2017**, *520*, 7–12. [CrossRef]
29. Suekuni, K.; Tsuruta, K.; Ariga, T.; Koyano, M. Variable-range-hopping conduction and low thermal conductivity in chalcogenide spinel Cu$_y$Fe$_4$Sn$_{12}$X$_{32}$ (X = S, Se). *J. Appl. Phys.* **2011**, *109*, 083709. [CrossRef]
30. Androulakis, J.; Do, J.-H.; Pcionek, R.; Kong, H.; D'Angelo, J.J.; Hogan, T.; Quarez, E.; Palchik, O.; Uher, C.; Short, J.; et al. Coexistence of Large Thermopower and Degenerate Doping in the Nanostructured Material Ag$_{0.85}$SnSb$_{1.15}$Te$_3$. *Chem. Mater.* **2006**, *18*, 4719–4721. [CrossRef]

Article

A Stable and Efficient Pt/n-Type Ge Schottky Contact That Uses Low-Cost Carbon Paste Interlayers

Pei-Te Lin [1], Jia-Wei Chang [2], Syuan-Ruei Chang [2], Zhong-Kai Li [2], Wei-Zhi Chen [2], Jui-Hsuan Huang [2], Yu-Zhen Ji [2], Wen-Jeng Hsueh [1,*] and Chun-Ying Huang [2,*]

[1] Photonics Group, Department of Engineering Science and Ocean Engineering, National Taiwan University, Taipei 10660, Taiwan; d09525014@ntu.edu.tw

[2] Department of Applied Materials and Optoelectronic Engineering, National Chi Nan University, Nantou 54561, Taiwan; james96068xd@gmail.com (J.-W.C.); s108328001@mail1.ncnu.edu.tw (S.-R.C.); s108328018@mail1.ncnu.edu.tw (Z.-K.L.); s108328036@mail1.ncnu.edu.tw (W.-Z.C.); g37833855@gmail.com (J.-H.H.); s108328031@mail1.ncnu.edu.tw (Y.-Z.J.)

* Correspondence: hsuehwj@ntu.edu.tw (W.-J.H.); cyhuang0103@ncnu.edu.tw (C.-Y.H.)

Abstract: Ge-based Schottky diodes find applications in high-speed devices. However, Fermi-level pinning is a major issue for the development of Ge-based diodes. This study fabricates a Pt/carbon paste (CP)/Ge Schottky diode using low-cost CP as an interlayer. The Schottky barrier height (Φ_B) is 0.65 eV for Pt/CP/n-Ge, which is a higher value than the value of 0.57 eV for conventional Pt/n-Ge. This demonstrates that the CP interlayer has a significant effect. The relevant junction mechanisms are illustrated using feasible energy level band diagrams. This strategy results in greater stability and enables a device to operate for more than 500 h under ambient conditions. This method realizes a highly stable Schottky contact for n-type Ge, which is an essential element of Ge-based high-speed electronics.

Keywords: carbon paste; n-type Ge; Schottky diodes; interlayer

Citation: Lin, P.-T.; Chang, J.-W.; Chang, S.-R.; Li, Z.-K.; Chen, W.-Z.; Huang, J.-H.; Ji, Y.-Z.; Hsueh, W.-J.; Huang, C.-Y. A Stable and Efficient Pt/n-Type Ge Schottky Contact That Uses Low-Cost Carbon Paste Interlayers. *Crystals* **2021**, *11*, 259. https://doi.org/10.3390/cryst11030259

Academic Editor: Michael Waltl

Received: 14 February 2021
Accepted: 4 March 2021
Published: 6 March 2021

Publisher's Note: MDPI stays neutral with regard to jurisdictional claims in published maps and institutional affiliations.

Copyright: © 2021 by the authors. Licensee MDPI, Basel, Switzerland. This article is an open access article distributed under the terms and conditions of the Creative Commons Attribution (CC BY) license (https://creativecommons.org/licenses/by/4.0/).

1. Introduction

Complementary metal oxide semiconductor (CMOS) technology is still the most common form of semiconductor device fabrication, which is capable of manufacturing sub-10 nm nodes using a traditional Si metal-oxide-semiconductor-field-effect-transistor (MOSFET) [1,2]. High-mobility channels are an effective booster. Germanium (Ge) is a possible alternative for Si for future high-speed applications because it features a high carrier mobility [3–5]. A good Schottky contact is an essential block for electronic circuits and devices. Generally, Ge-based Schottky diodes are affected by strong Fermi-level pinning [6,7]. Many studies try to modulate the Schottky barrier heights of metal/Ge junctions by inserting a thin insulator as an interfacial layer to minimize the effect of Fermi-level pinning [8–10]. Similarly, organic semiconductors that are used as the interfacial layer modify interface electronic states so the Schottky barrier height of the metal/Ge junction is decreased [11–13]. Recently, polymer poly (3,4-ethylenedioxythiophene)/poly (styrenesulfonate) (PEDOT/PSS) has been used for Schottky diodes as an interlayer, because it features high conductivity, solution processing capability, and is low-cost [14]. A. A. Kumar et al. demonstrated that using PEDOT/PSS as an interlayer for Pt/n-type Ge Schottky junctions increases the Schottky barrier height [13]. However, the semiconductor/PEDOT interface suffers from poor ambient stability because the hygroscopic nature of PSS means that water is absorbed, so it has a relatively weak tolerance to water [15]. Therefore, this obstacle limits its real application as an interlayer for high-speed Schottky diode.

Carbon is a cheap and accessible element on Earth [16]. Commercially available carbon paste (CP) is a low-cost electrical conductive printing ink with high conductivity and stability [16]. It contains conductive carbon particles and thermoplastic resins [17]. The

CP films are easily deposited using low temperature printing deposition processes, such as spin-coating, doctor-blading, inkjet-printing, and drop-casting [17,18]. The CP film is not easily oxidized after the sintering process, so it is resistant to high-temperature resistance and corrosion and is not subject to thermal shock [17,18]. Low-cost CP is widely found in applications in industry [17]. However, there are few related studies reporting a Schottky diode with a CP interlayer.

This study reports the fabrication of Pt/CP/n-Ge Schottky diodes. A low-cost CP is used as an interlayer to modify the metal/semiconductor interface. These diodes exhibit remarkable rectified performance, with a Schottky barrier height of 0.61 eV and a rectification ratio of 234 at ±1 V. These diodes also exhibit excellent long-term stability without encapsulation.

2. Materials and Methods

2.1. Materials

This study uses commercial CP from Alfa Aesar Co., Ltd. (Haverhill, MA, USA) (42465, alcohol-based) and n-type Ge (100) substrates, having a carrier concentration of 5.0×10^{15} cm^{-3}. The CP films were then spin-coated at 5000 rpm for 60 s, followed by pre-baking at 60 °C in an ambient atmosphere for 15 min on a hotplate. The sample was then sintered in a furnace in an Ar atmosphere at 300 °C for 30 min.

2.2. Characterization

The thickness of the CP film was ~70 nm, as measured using a profilometer. Atomic force microscopy (AFM) was used to determine the surface morphology of the CP films. The surface morphology of a CP film is quite smooth, as shown in Figure 1a. The root-mean-square (rms) roughness value for a CP film is 0.915 nm. Pt eletrodes (300 × 300 μm^2) were produced by means of sputtering through a shadow mask. The structure is shown in Figure 1a. For comparison, a reference device without a CP interlayer was fabricated using the same process conditions. The current–voltage (I–V) curves for the Schottky diodes were measured using a Keithley 2400 sourcemeter. The capacitance–voltage (C–V) plots for the Schottky diodes were recorded using an Agilent E4980A impedance analyzer.

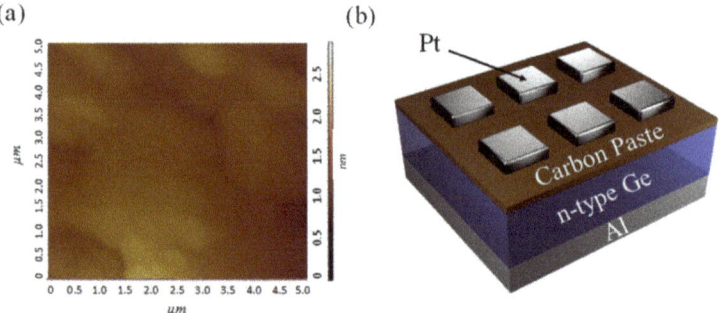

Figure 1. (a) Atomic force microscopy (AFM) image of a carbon paste (CP) film and (b) device structure of a Pt/CP/n-Ge Schottky diode.

3. Results and Discussion

3.1. I–V Characteristics

The effect of an interfacial layer on the electrical characteristics is considered. Figure 2 shows the experimental semi-logarithmic I–V characteristics for Pt/n-type Ge Schottky rectifiers with and without a CP interlayer in the dark. Both diodes exhibit excellent rectification behavior. The characteristic properties of the rectifying contact behavior mean that a reverse bias current has no significant effect on the voltage and there is an exponential increase in the forward bias current. A higher current rectification ratio is achieved for a

Pt/CP/n-Ge Schottky junction (234) than for its counterpart (20). Figure 2 shows that the leakage current in reverse bias for Pt/CP/n-Ge Schottky junction is less than that for the counterpart, which demonstrates that the CP interlayer creates an effective barrier between Pt and Ge.

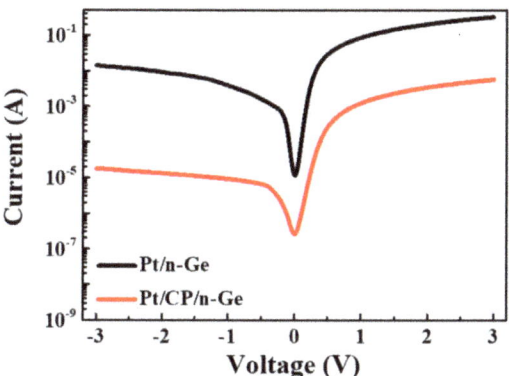

Figure 2. Experimental I–V curves for Pt/n-Ge Schottky diodes with and without a CP interlayer.

Besides, owing to the existence of bulk resistance of the CP interlayer, there is an obvious decrease at high current for the Pt/CP/n-Ge Schottky junction in the forward bias I–V plot. Standard thermionic emission theory is used to examine the electrical properties for a Schottky diode, which is given as follows [19]:

$$I = AA^*T^2 \exp(\frac{-q\Phi_B}{kT}) \left\{ \exp\left[\frac{q(V - IR_S)}{nkT}\right] - 1 \right\} \quad (1)$$

where A^* is the Richardson constant for Ge, which is 140 A cm^{-2}K^{-2}; $\Phi_{B,IV}$ is the barrier height; k is the Boltzmann constant; q is the electronic charge; R_s is the series resistance; and n is an ideality factor [12]. The extracted parameters for Schottky diodes are listed in Table 1. For Pt/n-Ge and Pt/CP/n-Ge Schottky junctions, $\Phi_{B,IV}$ and n are 0.57 eV and 1.08 and 0.65 eV and 1.95, respectively. The value of $\Phi_{B,IV}$ for the studied diode (0.65 eV) is higher than that for the counterpart (0.57 eV). The CP layer produces a higher barrier of 0.08 eV. This result is in agreement with the results of previous reports [11,13]. An organic interlayer prevents direct contact between the metal and Ge surface and significantly changes the interface states, even though the organic/inorganic interface is abrupt and unreactive [20–22]. A conventional Ge Schottky contact generally has a strong Fermi-level pinning effect, which leads to poor device performance (i.e., high n and low Φ_B). This result is attributed to the metal-induced gap states. Ion bombardment during plasma processing is another possible reason that the surface of the Ge substrate is destroyed, which induces defects and intermixing in the films. The values for n and Φ_B for the reference diode for this study (n = 1.08; Φ_B = 0.57) are comparable to those reported by previous studies [11–13]. Further improvement in the future is possible. For a conventional Schottky diode, the value for n is close to unity. This result shows that the diode current is mainly due to diffusion current and the pure thermionic emission theory fits well. However, the value for n for a Pt/CP/n-Ge Schottky junction is much higher than unity, possibly because of secondary mechanisms, such as CP of uneven thickness, series resistance, and a non-uniform distribution of dipoles due to the presence of an organic interfacial layer [23–26]. This phenomenon is also noted in previous studies that use an organic interfacial layer for Schottky diodes [11–13]. The exact reason for a higher n value is unclear. Generally, an interfacial layer is used to modulate the Schottky barrier height using Fermi level depinning. The value for Φ_B increases as the n value increases. The parameters for Pt/CP/n-Ge Schottky diodes could be further optimized.

Table 1. Schottky diode parameters derived using various methods. CP, carbon paste.

	I–V		dV/d(lnI) vs. I		H(I) vs. I		Norde		C–V
	Φ_B (eV)	n	R_S (Ω)	n	R_S (Ω)	Φ_B (eV)	R_S (Ω)	Φ_B (eV)	Φ_B (eV)
w/o CP	0.57	1.08	4.37	1.01	8.93	0.57	12.14	0.58	0.82
w/i CP	0.65	1.95	376	1.28	459	0.65	2735	0.72	0.9

The plots of Cheung's function $dV/d(\ln I)$ versus I and $H(I)$ versus I are used to accurately determine the Schottky parameters, $\Phi_{B,H(I)}$, n, and R_s, as shown in Figure 3. Cheung's function is expressed as follows [12,27]:

$$\frac{dV}{d(\ln I)} = \frac{nkT}{q} + IR_S \quad (2)$$

$$H(I) = V - \left(\frac{nkT}{q}\right)\ln\left(\frac{I}{AA^*T^2}\right) \quad (3)$$

$$H(I) = n\Phi_{B,H(I)} + IR_S \quad (4)$$

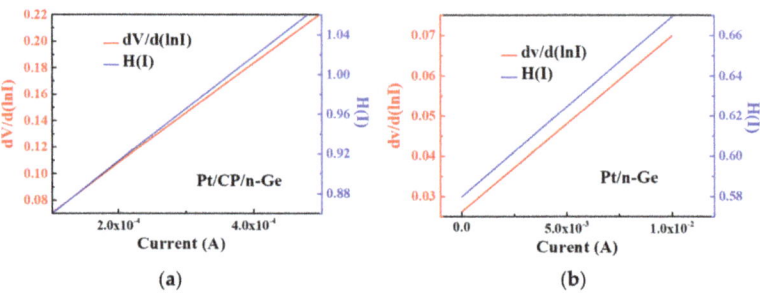

Figure 3. Plots of $dV/d(\ln I)$ versus I and $H(I)$ versus I using diode Equation (1) for Pt/n-Ge Schottky diodes (a) with and (b) without a CP interlayer.

The plot of $dV/d(\ln I)$ versus I is linear with a gradient of R_s and the y-intercept gives the ideality factor n. The respective values of R_s and n are 4.37 Ω and 1.01 for the Pt/n-Ge Schottky diode, and 376 Ω and 1.28 for the Pt/CP/n-Ge Schottky diode. The $H(I)$ versus I plot is a straight line and the y-axis intercept gives the value for $\Phi_{B,H(I)}$. The gradient of this plot can also be used to determine the value of R_s. From the $H(I)$ versus I plot, the R_s and $\Phi_{B,H(I)}$ values are 8.93 Ω and 0.57 eV for the Pt/n-Ge Schottky junction and 459 Ω and 0.65 eV for the Pt/CP/n-Ge Schottky junction. The large difference in the values is attributed to a deviation from the ideal model due to the insertion of the CP layer, so the value of R_s is higher [28]. The higher R_s limits the forward current, as shown in Figure 2. An additional decrease in voltage across the CP interlayer produces forward current conduction at a high voltage.

The values of $\Phi_{B,Norde}$ and R_s for Schottky junctions are also derived using the modified Norde function, which is expressed as follows [29]:

$$F(V) = \frac{V}{\gamma} - \frac{kT}{q}\ln\left(\frac{I(V)}{AA^*T^2}\right) \quad (5)$$

where $I(V)$ is the current obtained from the I-V curves and γ is an integer (dimensionless) that is greater than the value of n. Figure 4 presents the Norde plot as function of applied bias for Pt/n-Ge and Pt/CP/n-Ge Schottky junctions. The minimum point is used to obtain the corresponding voltage that is applied across the device. The value of $\Phi_{B,\,Norde}$ is derived using the following equation [29,30]:

$$\Phi_{B,Norde} = F(V_0) + \frac{V_0}{\gamma} - \frac{kT}{q} \tag{6}$$

where $F(V_0)$ is the minimum point of $F(V)$ and V_0 is the corresponding voltage. The value of R_s is determined using the following relation:

$$R_S = \frac{kT(\gamma - n)}{qI_0} \tag{7}$$

where I_0 is the minimum point of $F(V_0)$. The values for R_s and $\Phi_{B,\,Norde}$ are calculated as 12.14 Ω and 0.58 eV for the Pt/n-Ge Schottky junction and 2735 Ω and 0.72 eV for the Pt/CP/n-Ge Schottky junction, respectively. The R_s and $\Phi_{B,\,Norde}$ values that are derived using Norde's function are higher than those derived using Cheung's method because different fitting intervals are used for the I–V curve. Cheung's function focuses on the nonlinear region and the Norde plot considers the whole range of the I–V curve in forward bias [28,30]. Even if there is a difference in the values for different fitting techniques, Schottky contacts show reasonably good agreement in terms of the value of Φ_B.

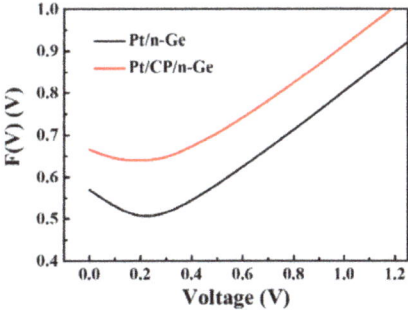

Figure 4. Norde plot for Pt/n-Ge Schottky diodes with and without a CP interlayer, using the diode Equation (1).

3.2. C–V Characteristics

The C–V characteristics of the Schottky diodes are shown in Figure 5. The capacitance of the depletion layer is described using a standard Mott–Schottky relationship [31]:

$$\frac{1}{C^2} = \frac{2(V_{bi} - V)}{A^2 q \varepsilon_s \varepsilon_0 N} \tag{8}$$

where A is the surface area, ε_s is the permittivity, ε_0 is the dielectric constant, V_{bi} is the built-in potential, and N is the carrier concentration. The x-intercept is V_{bi}. The gradient is N. The value of $\Phi_{B,\,CV}$ can be calculated as follows [31]:

$$\Phi_{B,CV} = \Phi_{bi} + \frac{kT}{q} \ln\left(\frac{N_C}{N}\right) \tag{9}$$

where N_c is the conduction band density states of 1.0×10^{-19} cm^{-3} [12]. A list of electronic parameters for the Schottky diodes is given in Table 1. The C–V characteristics give a barrier height of 0.90 eV for a Ge Schottky diode with a CP interlayer and 0.82 eV for a Pt/n-type Ge Schottky diode with no CP interlayer. The increase of $\Phi_{B,\,CV}$ is consistent with the results of I–V measurements when there is a CP interlayer. There is a relatively large discrepancy between the value for $\Phi_{B,\,IV}$ and $\Phi_{B,\,CV}$ because the conduction mechanism in these diodes does not exactly obey the thermionic emission theory. The values of $\Phi_{B,\,IV}$

derived from the *I–V* curve are sensitive to image force due to the current flow across the barrier [32]. The values of $\Phi_{B,CV}$ derived from the *C–V* curve are insensitive to potential fluctuations because the scale is much shorter than the space charge region [33]. The *C–V* measurement gives an average barrier height for the entire diode, but the current flows preferentially through the barrier minima during *I–V* measurement [33]. The difference between *C–V* and *I–V* measurement techniques results in different results for Schottky barrier height [34].

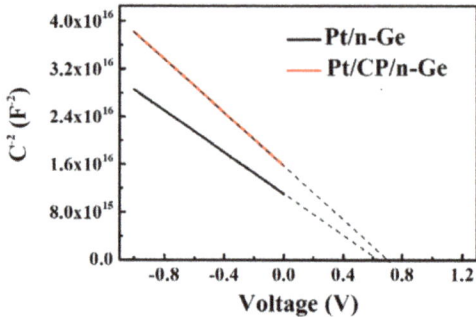

Figure 5. A^2/C^2–*V* characteristics for Pt/n-Ge Schottky diodes with and without a CP interlayer.

3.3. Energy Band Diagrams

The role of the CP interlayer for Pt/CP/n-Ge Schottky diodes is demonstrated using an energy diagram, as shown in Figure 6. At the metal/semiconductor interface, there are many dangling bonds, so there is a strong Fermi level pinning effect, as shown in Figure 6a. CP is an interlayer that conducts charged carriers with tiny resistance. The CP interlayer play a role of a dangling bond terminator at the Ge surface (Figure 6b).

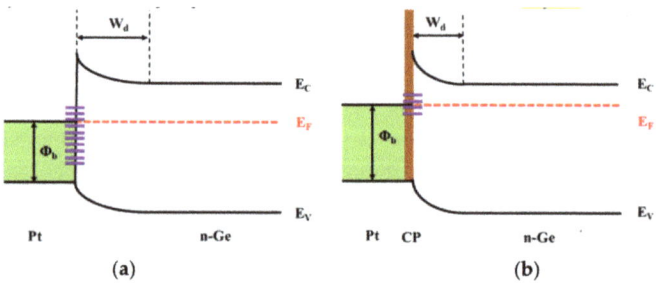

Figure 6. Energy band diagrams for Pt/n-Ge Schottky diodes (**a**) with and (**b**) without a CP interlayer.

3.4. Long-Term Reliability

Long-term reliability is a key factor for a diode for real applications [35–37]. To the best of the author's knowledge, there is no standard specification involving stability tests for Schottky diodes. To determine the long-term reliability of the CP interlayer, a Schottky diode with CP was driven under 0.1 A under ambient conditions, with no encapsulation or humidity control, over a period of 500 h. The fluctuation in Φ_B and *n* is negligible over the time of the test, as shown in Figure 7a,b. The organic layer is the only source for device degradation because the crystalline Ge and metals are relatively stable in air. Compared with the emerging organic layer for a Pt/n-Ge Schottky diode, such as PEDOT/PSS, the ambient stability for the device is poor because a PEDOT/PSS film contains water soluble ionic PSS [13]. The sheet resistance for pristine PEDOT/PSS film increases by more than 20% after storage in air for 1 week [38]. In addition, the carbon-containing diode exhibits long-term stability under a DC bias, which is attributed to the excellent thermal stability of CP. When current flows cross a contact, spot heat is generated in the constriction resistance [39], which decreases the performance of an organic semiconductor. CP is primarily composed of graphite powder and graphite is resistant to high temperatures [18]. Therefore, commercially available CP is eminently suited to the fabrication of highly stable and efficient Schottky diodes.

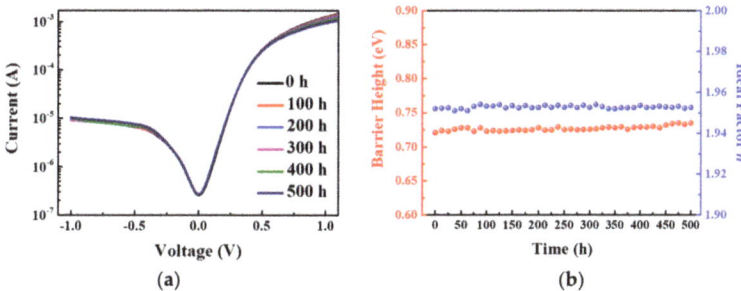

Figure 7. (a) *I–V* characteristics for a Pt/CP/n-Ge Schottky diode for different DC stress (0.1 A) times and (b) Φ_B and *n* as a function of stress time, as obtained from the diode equation (1): the measurements were performed at room temperature in ambient conditions.

4. Conclusions

An organic–inorganic structure is formed using a CP thin film as an interlayer for a Pt/n-Ge Schottky diode. To determine the electrical characteristics of the Pt/CP/n-Ge Schottky diodes, the *I–V* and *C–V* characteristics are measured at room temperature. This structure exhibits better rectifying behavior and results in a higher barrier height than a conventional Pt/n-Ge diode because the effective barrier height is increased when there is a CP interlayer. This study shows that CP is a reliable interlayer for an inorganic–organic hybrid Schottky diode.

Author Contributions: Conceptualization, P.-T.L.; methodology, J.-W.C, S.-R.C. and Z.-K.L.; formal analysis, W.-Z.C., J.-H.H. and Y.-Z.J.; writing—original draft preparation, P.-T.L.; writing—review and editing, P.-T.L.; supervision, W.-J.H. and C.-Y.H. All authors have read and agreed to the published version of the manuscript.

Funding: This research was funded by Ministry of Science and Technology of Taiwan under Contract No. MOST 106-2218-E-260-001-MY3.

Data Availability Statement: The data presented in this study are contained in this article.

Acknowledgments: This work is supported by the Ministry of Science and Technology of Taiwan under grant No. MOST 106-2218-E-260-001-MY3.

Conflicts of Interest: The authors declare no conflict of interest.

References

1. Chawanda, A.; Coelho, S.M.M.; Auret, F.D.; Mtangi, W.; Nyamhere, C.; Nel, J.M.; Diale, M. Effect of thermal treatment on the characteristics of iridium Schottky barrier diodes on n-Ge (100). *J. Alloy. Compd.* **2012**, *513*, 44–49. [CrossRef]
2. Ruan, D.-B.; Chang-Liao, K.-S.; Hong, Z.-Q.; Huang, J.; Yi, S.-H.; Liu, G.-T.; Chiu, P.-C.; Li, Y.-L. Radiation effects and reliability characteristics of Ge pMOSFETs. *Microelectron. Eng.* **2019**, *216*. [CrossRef]
3. Pfeiffer, U.R.; Mishra, C.; Rassel, R.M.; Pinkett, S.; Reynolds, S.K. Schottky Barrier Diode Circuits in Silicon for Future Millimeter-Wave and Terahertz Applications. *IEEE Trans. Microw. Theory Tech.* **2008**, *56*, 364–371. [CrossRef]
4. Lou, X.; Zhang, W.; Xie, Z.; Yang, L.; Yu, X.; Liu, Y.; Chang, H. Solution-processed high-k dielectrics for improving the performance of flexible intrinsic Ge nanowire transistors: Dielectrics screening, interface engineering and electrical properties. *J. Phys. D Appl. Phys.* **2019**, *52*. [CrossRef]
5. Khurelbaatar, Z.; Kil, Y.-H.; Shim, K.-H.; Cho, H.; Kim, M.-J.; Kim, Y.-T.; Choi, C.-J. Temperature Dependent Current Transport Mechanism in Graphene/Germanium Schottky Barrier Diode. *JSTS J. Semicond. Technol. Sci.* **2015**, *15*, 7–15. [CrossRef]

6. Chawanda, A.; Nyamhere, C.; Auret, F.D.; Mtangi, W.; Diale, M.; Nel, J.M. Thermal annealing behaviour of platinum, nickel and titanium Schottky barrier diodes on n-Ge (100). *J. Alloy. Compd.* **2010**, *492*, 649–655. [CrossRef]
7. Chen, Z.; Yuan, S.; Li, J.; Zhang, R. Thermal Stability Enhancement of NiGe Metal Source/Drain and Ge pMOSFETs by Dopant Segregation. *IEEE Trans. Electron Devices* **2019**, *66*, 5284–5288. [CrossRef]
8. Lee, H.-K.; Jyothi, I.; Janardhanam, V.; Shim, K.-H.; Yun, H.-J.; Lee, S.-N.; Hong, H.; Jeong, J.-C.; Choi, C.-J. Effects of Ta-oxide interlayer on the Schottky barrier parameters of Ni/n-type Ge Schottky barrier diode. *Microelectron. Eng.* **2016**, *163*, 26–31. [CrossRef]
9. Zhou, Y.; Ogawa, M.; Han, X.; Wang, K.L. Alleviation of Fermi-level pinning effect on metal/germanium interface by insertion of an ultrathin aluminum oxide. *Appl. Phys. Lett.* **2008**, *93*. [CrossRef]
10. Zhou, Y.; Han, W.; Wang, Y.; Xiu, F.; Zou, J.; Kawakami, R.K.; Wang, K.L. Investigating the origin of Fermi level pinning in Ge Schottky junctions using epitaxially grown ultrathin MgO films. *Appl. Phys. Lett.* **2010**, *96*. [CrossRef]
11. Kumar, A.A.; Reddy, V.R.; Janardhanam, V.; Seo, M.-W.; Hong, H.; Shin, K.-S.; Choi, C.-J. Electrical Properties of Pt/n-Ge Schottky Contact Modified Using Copper Phthalocyanine (CuPc) Interlayer. *J. Electrochem. Soc.* **2011**, *159*, H33–H37. [CrossRef]
12. Jyothi, I.; Janardhanam, V.; Rajagopal Reddy, V.; Choi, C.-J. Modified electrical characteristics of Pt/n-type Ge Schottky diode with a pyronine-B interlayer. *Superlattices Microstruct.* **2014**, *75*, 806–817. [CrossRef]
13. Ashok Kumar, A.; Rajagopal Reddy, V.; Janardhanam, V.; Yang, H.-D.; Yun, H.-J.; Choi, C.-J. Electrical properties of Pt/n-type Ge Schottky contact with PEDOT:PSS interlayer. *J. Alloy. Compd.* **2013**, *549*, 18–21. [CrossRef]
14. Enver Aydin, M.; Yakuphanoglu, F. Electrical characterization of inorganic-on-organic diode based InP and poly(3,4-ethylenedioxithiophene)/poly(styrenesulfonate) (PEDOT:PSS). *Microelectron. Reliab.* **2012**, *52*, 1350–1354. [CrossRef]
15. Huang, C.-Y.; Lin, P.-T.; Cheng, H.-C.; Lo, F.-C.; Lee, P.-S.; Huang, Y.-W.; Huang, Q.-Y.; Kuo, Y.-C.; Lin, S.-W.; Liu, Y.-R. Rectified Schottky diodes that use low-cost carbon paste/InGaZnO junctions. *Org. Electron.* **2019**, *68*, 212–217. [CrossRef]
16. Huang, C.-Y.; Lin, G.-Y.; Lin, P.-T.; Chen, J.-W.; Chen, C.-H.; Chien, F.S.-S. Influences of sintering temperature on low-cost carbon paste based counter electrodes for dye-sensitized solar cells. *Jpn. J. Appl. Phys.* **2017**, *56*. [CrossRef]
17. Gao, Y.; Chu, L.; Wu, M.; Wang, L.; Guo, W.; Ma, T. Improvement of adhesion of Pt-free counter electrodes for low-cost dye-sensitized solar cells. *J. Photochem. Photobiol. A Chem.* **2012**, *245*, 66–71. [CrossRef]
18. Mishraa, A.; Ahmadb, Z.; Zimmermannc, I.; Martineaud, D.; Shakoorb, R.A.; Touatia, F.; Riaze, K.; Al-Muhtasebf, S.A.; Nazeeruddinc, M.K. Effect of annealing temperature on the performance of printable carbon electrodes for perovskite solar cells. *Org. Electron.* **2019**, *65*, 375–380. [CrossRef]
19. Chasin, A.; Steudel, S.; Myny, K.; Nag, M.; Ke, T.-H.; Schols, S.; Genoe, J.; Gielen, G.; Heremans, P. High-performance a-In-Ga-Zn-O Schottky diode with oxygen-treated metal contacts. *Appl. Phys. Lett.* **2012**, *101*. [CrossRef]
20. Sreenu, K.; Venkata Prasad, C.; Rajagopal Reddy, V. Barrier Parameters and Current Transport Characteristics of Ti/p-InP Schottky Junction Modified Using Orange G (OG) Organic Interlayer. *J. Electron. Mater.* **2017**, *46*, 5746–5754. [CrossRef]
21. Tan, S.O.; Tecimer, H.; Cicek, O. Comparative Investigation on the Effects of Organic and Inorganic Interlayers in Au/n-GaAs Schottky Diodes. *IEEE Trans. Electron Devices* **2017**, *64*, 984–990. [CrossRef]
22. Tan, S.O. Comparison of Graphene and Zinc Dopant Materials for Organic Polymer Interfacial Layer Between Metal Semiconductor Structure. *IEEE Trans. Electron Devices* **2017**, *64*, 5121–5127. [CrossRef]
23. Winfried, M. Barrier heights of real Schottky contacts explained by metal-induced gap states and lateral inhomogeneities. *J. Vac. Sci. Technol. B* **1999**, *17*. [CrossRef]
24. Lonergan, M.C.; Jones, F.E. Calculation of transmission coefficients at nonideal semiconductor interfaces characterized by a spatial distribution of barrier heights. *J. Chem. Phys.* **2001**, *115*, 433. [CrossRef]
25. Gullu, O.; Cankaya, M.; Baris, O.; Biber, M.; Ozdemir, H.; Gulluce, M.; Turut, A. DNA-based organic-on-inorganic semiconductor Schottky structures. *Appl. Surf. Sci.* **2008**, *254*, 5175–5180. [CrossRef]
26. Gullu, O.; Aydogan, S.; Turut, A. High barrier Schottky diode with organic interlayer. *Solid State Commun.* **2012**, *152*, 381–385. [CrossRef]
27. Cheung, S.K.; Cheung, N.W. Extraction of Schottky diode parameters from forward current-voltage characteristics. *Appl. Phys. Lett.* **1986**, *49*, 85–87. [CrossRef]
28. Aubry, V.; Meyer, F. Schottky diodes with high series resistance: Limitations of forwardI-Vmethods. *J. Appl. Phys.* **1994**, *76*, 7973–7984. [CrossRef]
29. Norde, H. A modified forward I-V plot for Schottky diodes with high series resistance. *J. Appl. Phys.* **1979**, *50*, 5052–5053. [CrossRef]
30. Ocak, Y.S.; Guven, R.G.; Tombak, A.; Kilicoglu, T.; Guven, K.; Dogru, M. Barrier height enhancement of metal/semiconductor contact by an enzyme biofilm interlayer. *Philos. Mag.* **2013**, *93*, 2172–2181. [CrossRef]
31. Du, L.; Li, H.; Yan, L.; Zhang, J.; Xin, Q.; Wang, Q.; Song, A. Effects of substrate and anode metal annealing on InGaZnO Schottky diodes. *Appl. Phys. Lett.* **2017**, *110*. [CrossRef]
32. Xin, Q.; Yan, L.; Luo, Y.; Song, A. Study of breakdown voltage of indium-gallium-zinc-oxide-based Schottky diode. *Appl. Phys. Lett.* **2015**, *106*. [CrossRef]
33. Zhang, J.; Wang, H.; Wilson, J.; Ma, X.; Jin, J.; Song, A. Room Temperature Processed Ultrahigh-Frequency Indium-Gallium–Zinc-Oxide Schottky Diode. *IEEE Electron Device Lett.* **2016**, *37*, 389–392. [CrossRef]

34. Xin, Q.; Yan, L.; Du, L.; Zhang, J.; Luo, Y.; Wang, Q.; Song, A. Influence of sputtering conditions on room-temperature fabricated InGaZnO-based Schottky diodes. *Thin Solid Film.* **2016**, *616*, 569–572. [CrossRef]
35. Zou, C.; Huang, C.Y.; Sanehira, E.M.; Luther, J.M.; Lin, L.Y. Highly stable cesium lead iodide perovskite quantum dot light-emitting diodes. *Nanotechnology* **2017**, *28*, 455201. [CrossRef] [PubMed]
36. Huang, C.-Y.; Zou, C.; Mao, C.; Corp, K.L.; Yao, Y.-C.; Lee, Y.-J.; Schlenker, C.W.; Jen, A.K.Y.; Lin, L.Y. CsPbBr3 Perovskite Quantum Dot Vertical Cavity Lasers with Low Threshold and High Stability. *ACS Photonics* **2017**, *4*, 2281–2289. [CrossRef]
37. Qian, H.; Wu, C.; Lu, H.; Xu, W.; Zhou, D.; Ren, F.; Chen, D.; Zhang, R.; Zheng, Y. Bias stress instability involving subgap state transitions in a-IGZO Schottky barrier diodes. *J. Phys. D Appl. Phys.* **2016**, *49*. [CrossRef]
38. Kim, S.; Sanyoto, B.; Park, W.-T.; Kim, S.; Mandal, S.; Lim, J.-C.; Noh, Y.-Y.; Kim, J.-H. Purification of PEDOT:PSS by Ultrafiltration for Highly Conductive Transparent Electrode of All-Printed Organic Devices. *Adv. Mater.* **2016**, *46*. [CrossRef]
39. Williamson, J.B.P.; Allen, N. Thermal stability in graphite contacts. *Wear* **1982**, *78*, 38–49. [CrossRef]

Article

The Impact of Interfacial Charge Trapping on the Reproducibility of Measurements of Silicon Carbide MOSFET Device Parameters

Maximilian W. Feil [1,2,*], Andreas Huerner [2], Katja Puschkarsky [1,2], Christian Schleich [3], Thomas Aichinger [4], Wolfgang Gustin [2], Hans Reisinger [2] and Tibor Grasser [1]

1. Institute for Microelectronics, TU Wien, 1040 Wien, Austria; Katja.Waschneck@infineon.com (K.P.); grasser@iue.tuwien.ac.at (T.G.)
2. Infineon Technologies AG, 85579 Neubiberg, Germany; Andreas.Huerner@infineon.com (A.H.); Wolfgang.Gustin@infineon.com (W.G.); Reisinger.external4@infineon.com (H.R.)
3. CDL for Single-Defect Spectroscopy at the Institute for Microelectronics, TU Wien, 1040 Wien, Austria; schleich@iue.tuwien.ac.at
4. Infineon Technologies Austria AG, 9500 Villach, Austria; Thomas.Aichinger@infineon.com
* Correspondence: Maximilian.Feil@infineon.com; Tel.: +49-892-343-9095

Received: 6 November 2020; Accepted: 11 December 2020; Published: 16 December 2020

Abstract: Silicon carbide is an emerging material in the field of wide band gap semiconductor devices. Due to its high critical breakdown field and high thermal conductance, silicon carbide MOSFET devices are predestined for high-power applications. The concentration of defects with short capture and emission time constants is higher than in silicon technologies by orders of magnitude which introduces threshold voltage dynamics in the volt regime even on very short time scales. Measurements are heavily affected by timing of readouts and the applied gate voltage before and during the measurement. As a consequence, device parameter determination is not as reproducible as in the case of silicon technologies. Consequent challenges for engineers and researchers to measure device parameters have to be evaluated. In this study, we show how the threshold voltage of planar and trench silicon carbide MOSFET devices of several manufacturers react on short gate pulses of different lengths and voltages and how they influence the outcome of application-relevant pulsed current-voltage characteristics. Measurements are performed via a feedback loop allowing in-situ tracking of the threshold voltage with a measurement delay time of only 1 µs. Device preconditioning, recently suggested to enable reproducible BTI measurements, is investigated in the context of device parameter determination by varying the voltage and the length of the preconditioning pulse.

Keywords: hysteresis; device parameters; reproducibility; device characteristics; silicon carbide; threshold voltage; current-voltage characteristics; IV-curve; preconditioning

1. Introduction

Silicon carbide (SiC) is a promising wide band gap semiconductor material. Especially in power applications, 4H-SiC metal-oxide-semiconductor field-effect transistors (MOSFETs) exhibit advantages over comparable silicon (Si) technologies. Owing to its wide band gap of around 3.26 eV [1], the critical breakdown field of SiC power MOSFETs is around ten times higher than the one of Si devices [2]. Hence, SiC devices of the same voltage and on-state resistance class can be made considerably smaller than comparable Si-based devices, leading to reduced device capacitances and enabling higher switching frequencies at lower losses [3,4]. This dramatically reduces volume and weight of inductors, filter capacitors, cooling components and hence, total system costs. Operation at higher frequency, more flexible thermal capability, and robustness to hard commutation events make SiC MOSFETs particularly suited for high efficiency topologies and high density designs. The wide band gap nature of the semiconductor further allows device operation at very high operating temperatures [5], which makes SiC MOSFETs ideal candidates for harsh environments.

SiC power MOSFETs have already been manufactured by semiconductor companies for several years. Commercially available designs differ with regard to the crystal direction along the inversion channel. Vertical power MOSFETs, which employ a trench design, allow for a higher cell density on the wafer, show a higher channel mobility, but the manufacturing process is more complex. In contrast, their planar counterparts exhibit a lower channel mobility, as the crystal direction differs, the cell density is lower, but manufacturing is less complex [6,7]. Regarding device performance, SiC trench MOSFETs have the potential to show even lower switching and conduction losses than planar designs, leading to an increase in efficiency of applications [8].

Although the experience in manufacturing SiC devices has been growing continuously in recent years, device reliability is still one of the most important aspects under focus of research and development. Besides extrinsic defects in the gate oxide [9], bias temperature instability (BTI) is a topic that requires particular attention [10]. An applied gate voltage, that can be either positive or negative, leads to charge capture events in defects inside the SiO_2 gate insulator. These defects are either located deeper in the oxide, in the interfacial transition region, or directly at the interface between SiC and SiO_2. Such trapping and detrapping events lead to a transient shift of the threshold voltage and can be described by first-order reactions [11]. A peculiarity of SiC MOSFETs compared to Si counterparts is the high concentration of defects with short capture and emission time constants of roughly 10^{12} cm^{-2} [12,13]. These defects transform the threshold voltage to a rather dynamic quantity under application conditions [14]. In contrast to Si devices, in which threshold voltage shifts appear only as a gradual long-term drift (classical BTI), the threshold voltage of SiC devices shows a very fast transient reaction to even very short gate pulses in the submicrosecond range.

Two of the most important parameters of power MOSFETs are the threshold voltage V_{th} and the on-state resistance $R_{DS,on}$. For Si based MOSFETs, it is common practice to measure these parameters with an accuracy and reproducibility of 10 mV and 1%, respectively. Most importantly, application- or reliability-engineers can easily check or reproduce the corresponding datasheet values. For SiC MOSFETs, measuring the threshold voltage and the on-state resistance is more complex. The short-term threshold voltage dynamics discussed above can be on the order of a few volts and arise even at very low gate voltages within very short times. These transient threshold voltage shifts have a negative influence on the reproducibility and comparability of measurements. New measurement procedures and their standardization, involving device preconditioning before readouts, are currently discussed [15–17].

At the moment, dedicated studies on the impact of the short-term dynamics of interfacial charge trapping on the reproducibility of device parameter measurements and investigations on differences between preconditioning pulses do not exist. In this study, we therefore characterize four different state-of-the-art commercially available SiC power MOSFETs from different vendors with regard to the charge trapping dynamics during current-voltage characteristics (IV-curve) measurements. It will be shown how these transient threshold voltage shifts quantitatively enter device parameters. Furthermore, we will examine the impact of preconditioning pulse length and bias variations on the reproducibility of threshold voltage readouts.

2. Materials and Methods

The employed commercially available SiC power MOSFETs are from four different vendors. For reasons of comparability, the MOSFETs have the same maximum voltage rating of 1200 V and comparable maximum gate-source voltage ratings. Furthermore, the typical on-state resistance is comparable and in the range of 160–500 mΩ at room temperature. The devices are arbitrarily labeled as A, B, C, and D, whereby A and B have a trench design and C and D are planar MOSFETs. Additional external gate resistors in series to the respective positive or negative voltage source enable defined transitions between two voltage levels of different polarity. The gate resistors are chosen to set the rise and fall times for a transition between −10 V and 20 V to 50 ns for all devices, which is typical for fast switching applications of 100 kHz. Thereby, overshoots and undershoots are avoided. An overview of the different samples is found in Table 1.

Table 1. Overview of selected datasheet values and properties of the four different devices. The shown quantities are the maximum drain-source voltage $V_{(BR)DSS}$, the threshold voltage V_{th}, the minimum and maximum gate-source voltage ratings $V_{GS,min}$ and $V_{GS,max}$, the input capacitance C_{iss}, the design type, and a figure of merit.

Label	$V_{(BR)DSS}$ [V]	V_{th} [V]	$V_{GS,min}$ [V]	$V_{GS,max}$ [V]	C_{iss} [pF]	Design	$(C_{iss} R_{DS,on})^{-1}$ [ns^{-1}]
A	1200	3.5–5.7	−7	23	182	trench	16
B	1200	2.7–5.6	−4	22	398	trench	16
C	1200	1.8–N/A	−10	25	290	planar	7
D	1200	2.0–4.0	−10	25	259	planar	14

All measurements are performed at room temperature with a homebuilt setup enabling excellent control over the performed measurement procedures and ultra-fast threshold voltage readouts [13,18], which are executed by forcing a drain-source current of 1 mA at a drain-source voltage of 1 V and by tracking the corresponding gate voltage via an operational amplifier based feedback loop. The setup allows measurement delay times down to 1 µs.

3. Results

In this study, we define the threshold voltage as the gate voltage that is necessary to yield a certain drain-source current at a defined drain-source voltage. This allows its measurement by continuously forcing the chosen drain-source current via a feedback loop. As shown in Figure 1a, the resulting continuous application of the threshold voltage during a measurement leads to considerable shifts on the order of several hundred millivolts. The striking consequence is that charge trapping is present even at the comparably low threshold voltage that hence changes over time. Naturally, such effects also affect the drain-source current because a shift in the threshold voltage changes the gate voltage overdrive. This is especially visible in SiC MOSFETs when switching from negative to positive gate voltages (see Figure 1b). A similar effect can hardly be observed in Si devices (see inset Figure 1b).

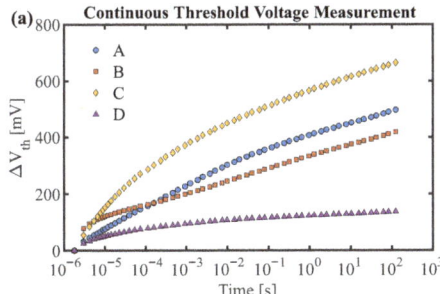
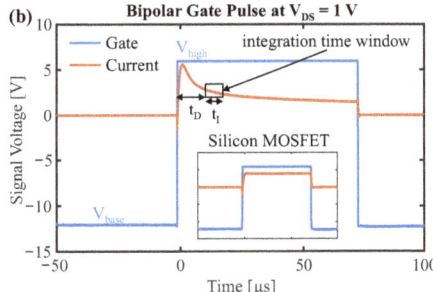

Figure 1. (a) The shift of the threshold voltage during a measurement over time of all devices A–D. Note that the threshold voltage is continuously applied to the gate (constant $I_{DS} = 1$ mA) triggering electron trapping. (b) Gate signal of a 70 µs pulse from $V_{base} = -12$ V to $V_{high} = 6$ V and a signal proportional to the corresponding drain-source current. Arrows indicate the measurement delay time t_D and the integration time t_I exemplarily for one out of several readouts. The current signal decreases monotonously during the pulse. The inset shows the same for a silicon MOSFET for comparison, where this effect is negligible.

As such drain-source current variations will of course arise during pulsed IV-curve measurements and thus influence the measured device parameters, we will investigate in the following the transient threshold voltage shift after short gate pulses and the drain-source current variations during pulsed IV-curves and finally correlate these two effects.

3.1. Threshold Voltage Dynamics after Short Gate Pulses

First, we determined the impact of short gate pulses of both positive and negative polarity on the threshold voltage. The measurement scheme is shown in the inset of Figure 2a. After a pulse of a certain length t_{pulse} and voltage $V_{GS,pulse}$ had been applied to the gate, while keeping both drain and source contact grounded, the threshold voltage transients were measured for around 2 min via the feedback loop. Remember that during the recovery phase the threshold voltage is continuously applied to the gate. Finally, the shift of the threshold voltage is determined by comparing the value at a specific recovery time to the threshold voltage at the same recovery time in a measurement where no pulse had been applied before (see Figure 1a). Each measurement sequence, consisting of pulses of different lengths and voltages, is performed with the same device, whereby the single measurements are separated by additional recovery periods of 2 min with all terminals grounded. These additional recovery periods allow the device to return to its pristine state. Exemplary recovery traces from the described measurements are presented in Figure 2 for each device. All plots contain the threshold voltage shift recovery traces after positive and negative pulses of different lengths.

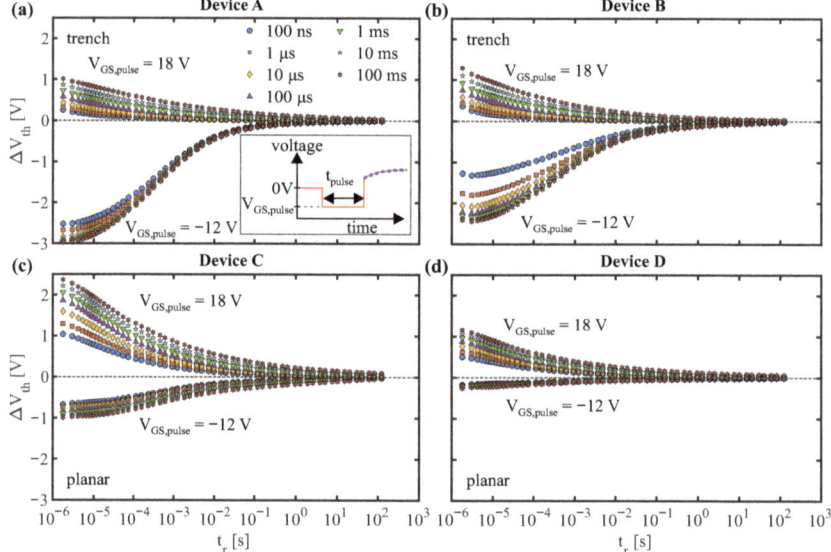

Figure 2. The recovery traces of the threshold voltage shift, respectively after a positive 18 V ($\Delta V_{th} > 0$) or a negative -12 V ($\Delta V_{th} < 0$) gate pulse. The pulse length t_{pulse} was varied between 100 ns and 100 ms. The subfigures (**a**–**d**) correspond to the respective devices A–D. The inset in (**a**) illustrates the measurement scheme consisting of an exemplary negative pulse followed by a continuous measurement of the threshold voltage.

Based on these results, Figure 3 illustrates the dependence of the threshold voltage shift on the pulse length for the two shown pulse voltages, while keeping the recovery time constant at 1.8 µs. Obviously, the longer the pulse length, the higher the absolute threshold voltage shift. However, note that for the negative voltage, the t_{pulse}-dependence above 1 ms is rather weak. Regarding the design impact, the two planar devices show less transient threshold voltage shift after the negative pulse than the two trench devices. In contrast, the two trench devices vastly show less threshold voltage shift for the positive pulse, whereby the maximum transient shift is observed in the planar device C. Note that pulses of only 100 ns lead to a shift of the threshold voltage in the volt regime.

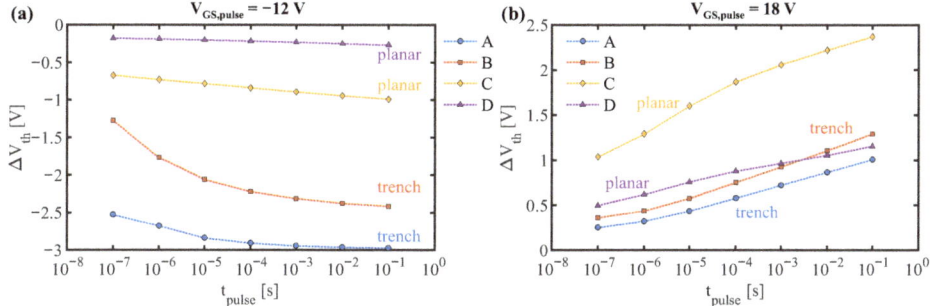

Figure 3. The pulse length dependence of the threshold voltage shift at the shortest measured recovery time of 1.8 µs of the tested devices A–D for the two pulse voltages of (**a**) −12 V and (**b**) 18 V. The full data set is shown in Figure 2.

The same measurements, as presented in Figure 2, were performed for several other voltages yielding the pulse voltage dependence of the threshold voltage shift for the two pulse lengths 1 µs and 100 ms respectively (see Figure 4). For positive pulse voltages, the behavior is similar for all devices. Regarding negative pulse voltages, the trench devices show a stronger pulse voltage dependence than the planar devices.

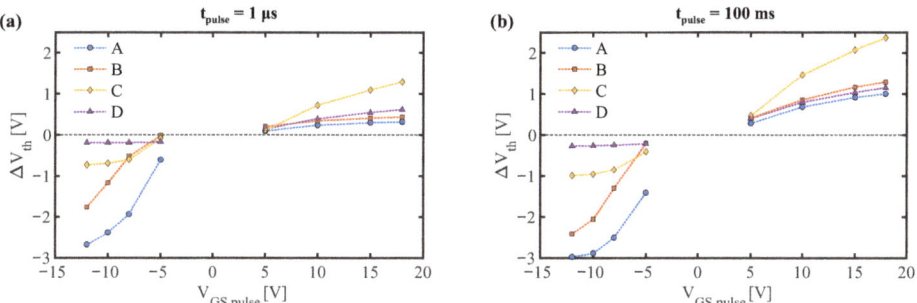

Figure 4. The pulse voltage dependence of the threshold voltage shift at the shortest measured recovery time of 1.8 µs of the tested devices A–D for the two pulse lengths (**a**) 1 µs and (**b**) 100 ms. The full data set is shown in Figure 2.

An obvious question is how long one has to wait after a pulse to come back to the original threshold voltage within a certain limit, which is shown in Figure 5. For a pulse length of 1 µs and negative pulse voltages, the time can reach up to the order of 1–100 s at which there is still a remaining threshold voltage shift of 50 mV, even though the causing pulse length was several orders of magnitude shorter. For the highest positive pulse voltage, the time to reach the limit is lower. In comparison to longer pulses of 100 ms and a higher precision limit of 200 mV, the times are roughly on the same order. Interestingly, the tested trench devices reach the respective precision level for positive voltages faster than the planar devices.

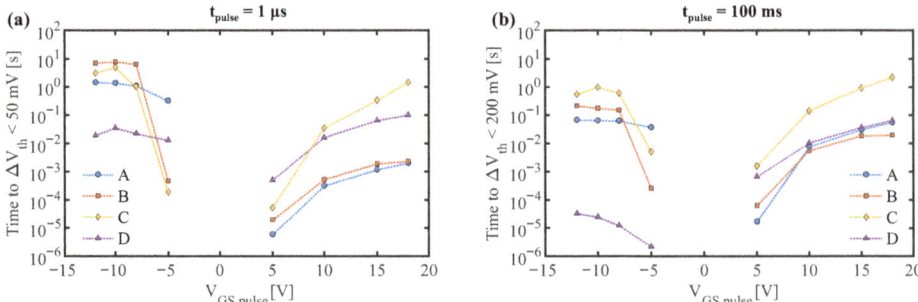

Figure 5. The recovery time necessary to reach a threshold voltage shift lower than a certain limit in dependence on the pulse voltage for a pulse length of (**a**) 1 µs, a limit of 50 mV and for a pulse length of (**b**) 100 ms and a limit of 200 mV. The full data set is shown in Figure 2.

3.2. Drain-Source Current Voltage Characteristics and On-State Resistance

In this section, we study the impact of interfacial charge trapping on IV-curves and device parameters. A short gate pulse of 70 µs typically leads to a dissipated power $P = V_{DS} \cdot I_{DS} = 3\,\text{W}$ resulting in a junction temperature increase of only 1 °C. In order to avoid self-heating, all IV-curves are hence executed in a pulsed mode with sufficiently long off-state times. Commercial curve tracers pulse the drain-voltage while slowly sweeping the gate voltage. Although this technique works perfectly for Si-MOSFETs or IGBTs, it is devastating for SiC MOSFETs because in their case, this method would lead to a drastic increase in the threshold voltage during the sweep by more than 1 V (see Figure 4). The gate overdrive $V_{GS} - V_{th}$ would thus decrease. Hence, for SiC MOSFETs pulsing the drain bias has to be replaced by pulsing the gate. A long negative pulse at a base voltage has to be applied between the positive gate pulses in order to discharge all traps that had been charged during the preceding positive pulse and bring the electrically active defects in a defined charge state. Furthermore, pulsed IV-curves can be adapted to typical application conditions of SiC MOSFETs which are characterized by a pulsed gate-source voltage up to 100 kHz. Thus, they provide application-relevant drain-source current measurement values.

Recalling Figure 1b, the measured drain-source current in pulsed IV-curves with a fixed drain-source voltage is influenced by four parameters: the base voltage V_{base}, the length of the base pulse, the measurement delay t_D, and the integration time t_I. As the integration time can be interpreted as an average over different delay times, we do not consider its dependence here and choose a rather small integration time of 3.6 µs compared to the pulse length of 70 µs. The length of the base pulse is also kept constant at 1 ms because the threshold voltage shift after negative gate pulses is almost identical for all measured pulse lengths longer than 1 ms (see Figure 3a).

For a precise characterization of the drain-source current variations during pulsed IV-curves, we measured a series of them, whereby the base voltage V_{base} was varied and the current was measured after different delay times. Figure 6 shows the difference in the IV-curve, the on-state resistance, and the transconductance for two different base voltages V_{base} but at the same delay time. Obviously, a lower base voltage leads to an increased drain-source current. Deviations from a parallel shift of the IV-curve hint towards changes in the channel mobility. Though the influence of the base voltage is minimized for device D, as could be expected from low hole trapping at negative gate voltage (see Figure 4a), all devices are affected. This is in accordance with observations in previous literature [19]. The reduction in the on-state resistance reaches up to 20 mΩ ($\approx$4%) for device A and even 34 mΩ ($\approx$8%) for device C.

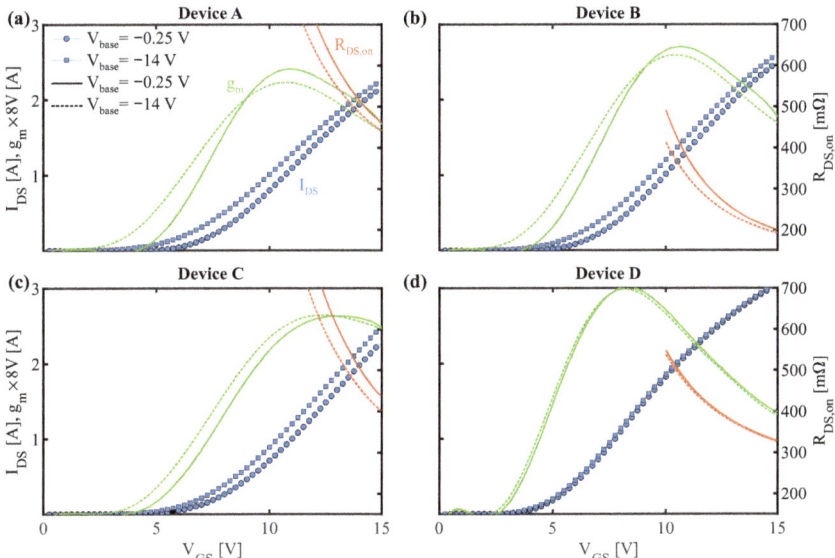

Figure 6. The drain-source current I_{DS}, transconductance g_m and on-state resistance $R_{DS,on}$ for a measurement delay of 7 µs and two different base voltages V_{base} with an off-state time of 1 ms. The subfigures (**a**–**d**) correspond to the respective devices A–D. The drain-source voltage V_{DS} was set to 1 V, except for device B ($V_{DS} = 0.5$ V). A lower base voltage yields a higher drain-source current.

An illustration of the impact of the delay time on the drain-source current is shown in Figure 7, where two different delay times were used with the same base voltage. The observed change in the drain-source current is on the same order as the one that resulted from a change of the base voltage. Again, while the influence is minimized for device D, all devices show a significant change. Such a small difference in delay time of only 60 µs already translates for device A into a shift of the maximum transconductance of a few hundred millivolt and a difference in the on-state resistance of up to 20 mΩ (≈5%) at a gate bias of 15 V. For device C, the difference even reaches 30 mΩ (≈8%).

For more insights, a detailed dependence of the change in the drain-source current on the base voltage is shown in Figure A2 for different delay times. We define the quantities $\Delta I_{DS}(t_D)$ and $\Delta I_{DS}(V_{base})$, as illustrated in Figure A2, to be the maximum observed change in the drain-source current due to a change in the delay time (relative to the lowest delay time) or the base voltage (relative to the lowest base voltage). The quantity $\Delta I_{DS}(V_{base})$ of the tested devices ranges from 0.4% to 8.7%. With decreasing base voltage, the drain-source current increases monotonously for all devices except for device C. Here, the drain-source current shows a local minimum at -2 V. In this case, the evaluation relative to the minimum would lead to a variation in the drain-source current of over 10%. Due to its generally small current variations, device D is the only device showing noise in the change of the drain-source current. At a certain base voltage, that is different for each of the four tested devices, the change in the drain-source current saturates. The quantity $\Delta I_{DS}(t_D)$ is different for each device and ranges from 0.2% to 2.5%.

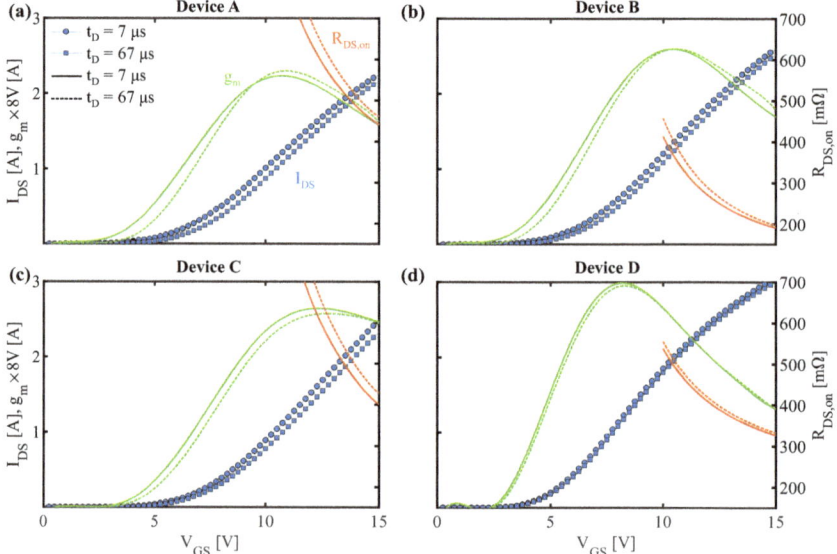

Figure 7. The drain-source current I_{DS}, transconductance g_m, and on-state resistance $R_{DS,on}$ for a base voltage of of $-14\,\mathrm{V}$ and two different delay times t_D with an off-state time of 1 ms. The subfigures (**a**–**d**) correspond to the respective devices A–D. The drain-source voltage V_{DS} was set to 1 V, except for device B ($V_{DS} = 0.5\,\mathrm{V}$). A longer delay time yields a lower drain-source current.

As observed in the measurements in Figure 2, the lower the pulse voltage and the longer the pulse, the larger is the absolute change of the threshold voltage and, thus, the larger is the change in the gate voltage overdrive leading to a higher drain-source current. As a result, the change in drain-source current correlates with the transient threshold voltage shift. In order to prove this correlation, drain-source current measurements with different delay times were plotted versus the observed change in the threshold voltage after a negative pulse that corresponds exactly to the off-state time and base voltage in the drain-source current measurement. The results are presented in the appendix in Figure A3. Pearson correlation coefficients indicate strong linear correlation near unity.

3.3. Device Preconditioning

Device preconditioning is often performed with either a negative accumulation pulse or a positive inversion pulse prior to drain-source current or threshold voltage measurements. Sometimes, both types of pulses are used consecutively. As discussed in the previous sections, any gate pulse, even the measurement itself, will cause an out-of-equilibrium state of the interfacial charge traps affecting the threshold voltage. For device reliability, stress tests where a threshold voltage shift is determined relative to a reference readout before applied device stress, it is sufficient to be able to create at each readout, including the reference readout, a defined out-of-equilibrium state. This is a very good approach as long as the timing and the voltages of the preconditioned measurement scheme are precisely kept constant and as long as the underlying degradation mechanism does not affect the transient threshold voltage shift [16,17]. However, an inaccurate recovery time between the end of the preconditioning pulse and the readout can have an impact on the precision and the reproducibility of the measurement. In cases where a defined recovery time cannot be provided, a close to equilibrium charge state of the involved traps has to be achieved before each measurement. In the following, we investigate this case with a negative accumulation pulse (see inset of Figure 8a). The idea is to remove trapped electrons with short time constants which can help to achieve reproducible measurements and to measure mainly application-relevant permanent drift components in long-term stress tests of positive gate bias. For this study, we chose a stress pulse of 1 s at 20 V. Such a stress pulse could typically arise during measurements of IV-curves or as an unintentional variation of the applied gate voltage before a threshold voltage readout in reliability tests. In order to reveal the impact of preconditioning pulses on the removal of the captured electrons, we executed measurements where preconditioning pulse voltage and length were varied. As the threshold voltage of device A is more sensitive to negative gate voltages than the one of the other devices, we limited the measurements to this device. Figure 8 shows the threshold voltage traces after stress and preconditioning. Each plot corresponds to one of four preconditioning voltages V_{prec}, whereby the pulse length t_{prec} was varied. The stress without preconditioning causes a transient shift of the threshold voltage of approximately 1 V. At a preconditioning voltage of -3 V, pulse lengths between 1 and 100 µs reduce the transient shift already significantly. At a pulse length of 1 ms, an initially positive slope of the recovery trace indicates the emission of holes. However, for longer recovery times, the recovery is still dominated by the emission of electrons. The preconditioning pulse is already sufficient to change the sign of the threshold voltage shift, although the difference to the stress pulse is three orders of magnitude in time and the preconditioning voltage is relatively small. At a lower preconditioning voltage of -6 V, the threshold voltage shift is not only negative for all measured preconditioning pulse lengths, but the absolute value of the threshold voltage shift is even increased compared to the case of no preconditioning. Decreasing the preconditioning voltage even more results in gradually increasing absolute shifts of the threshold voltage, whereby the preconditioning pulse length looses its influence, as the different recovery traces merge.

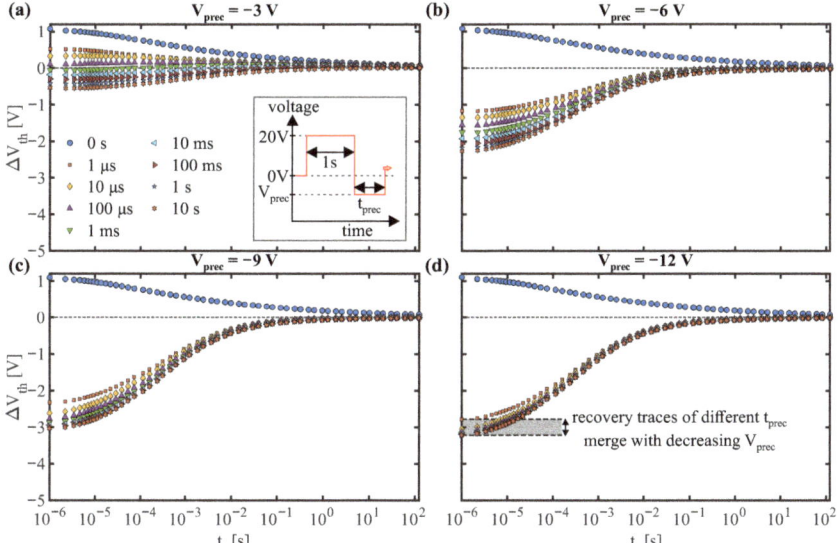

Figure 8. The recovery traces of the threshold voltage shift without preconditioning (0 s) and with preconditioning pulses of lengths between 1 µs and 10 s measured with device A. The subfigures (**a**–**d**) correspond to different preconditioning voltages V_{prec} between -3 V and -12 V. The inset in (**a**) illustrates the measurement scheme consisting of the stress pulse (20 V for 1 s) followed by the preconditiong pulse. Except for the case of the -3 V preconditioning with pulse lengths between 1 µs and 100 µs, all preconditioning pulses are strong enough to change the sign of the threshold voltage shift.

This merging becomes particularly visible in Figure 9. It shows the dependence of the recovery time needed after preconditioning to return to a threshold voltage shift below 100 mV. To remove the influence on the threshold voltage of the applied positive pulse, a preconditioning pulse of

−3 V for 1 ms appears to be a very good compromise. However, the impact on the needed recovery time of a small error in preconditioning time is considerably high at 1 ms, which would lead to poor reproducibility. For the lower preconditioning voltages, the required recovery time is almost the same, especially for longer preconditioning pulses. The necessary additional recovery time after preconditioning is around 440 ms. Without preconditioning, the required recovery time is approximately 20–30 s. Negative preconditioning can hence reduce the recovery time required to come back to the initial threshold voltage by around two orders of magnitude.

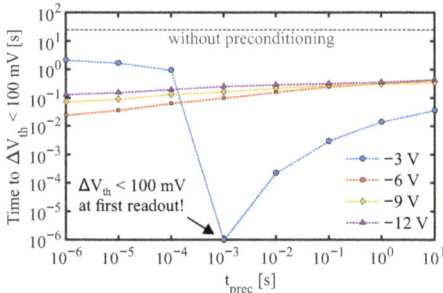

Figure 9. Required recovery time after preconditioning to return to a threshold voltage shift below 100 mV as a function of the preconditioning time. The raw measurement data is shown in Figure 8.

4. Discussion

Due to the extraordinarily wide distribution of capture time constants of defects in SiC devices, all presented measurements are heavily affected by charge trapping. Continuously measuring the threshold voltage via a feedback loop reveals the drastic time dependence of such measurements, during which the applied threshold voltage leads to electron capture events in the electrically active defects, resulting in return in a dynamic increase of the threshold voltage. Short-term charge trapping is inherent in all measurements, as it cannot be avoided due to measurement delays. The observed asymmetry between capture and emission times has already been investigated by means of capture and emission time maps [13], confirming the existence of defects with longer emission than capture time constants. The observation of hysteresis in swept IV-curves is a direct implication of this effect [12]. Furthermore, fast gate voltage up sweeps starting at different negative values are also affected (see Figure A1). As shown in Figure 1b, switching a SiC MOSFET from a negative to a positive gate bias results in a gradually decreasing drain-source current. This effect can be linked to charge trapping via the observed threshold voltage shift. During the time at negative gate bias, fast hole trapping, probably at the interface [12,20], leads to a negative shift of the threshold voltage (see Figure 2). This leads to an initially increased gate voltage overdrive and hence to a higher drain-source current. During the positive gate pulse, the trapped holes gradually get emitted which increases the threshold voltage again. As a result, the gate voltage overdrive and the drain-source current decrease. A strong linear correlation between the drain-source current and the threshold voltage shift undermines this mechanism (see Figure A3). Small deviations can be attributed to variations in the channel mobility and the influence of the difference in the positive gate voltage during threshold voltage measurement and drain-source current measurement.

This short-term response of the threshold voltage leads to difficulties regarding the reproducibility of device parameter measurements. At a high level of 15 V at the gate (see Figure A2), a modification of the base voltage can lead to a change in the drain-source current of almost 10%. Furthermore, an increase in delay time of only 60 µs can reduce the drain-current by up to 2.5%. These variations propagate to other device parameters, such as transconductance and on-state resistance (see Figures 6 and 7). Precise device parameter measurements have to take such variations into account and they have to be considered when comparing results from different measurement schemes or even different measurement setups. However, short-term threshold voltage shifts should not necessarily be considered as a detrimental property in most applications because these transient effects are fully reversible and therefore do not cause typical BTI-induced reliability issues.

Differences between tested devices can be explained by multiple factors. The crystal plane of the interface and manufacturing processes, such as post oxidation anneals, influence the atomic structure of the interface and hence the trapping dynamics of charge carriers.

In device characterization measurements or device reliability tests, it is often required to remove electrons with small emission time constants trapped during positive gate stress. The application of a negative preconditioning pulse prior to threshold voltage readouts does not only accelerate the emission of trapped electrons, but it also triggers the capture of holes. On short time scales, the impact of the negative preconditioning pulse quickly dominates the recovery behavior (see Figure 8). For our exemplary stress pulse, the emission of most electrons required up to 440 ms to get the absolute threshold voltage shift back below 100 mV resulting in an acceleration of recovery by around two orders of magnitude. Below -6 V, this was almost independent of the considered preconditioning pulse length and of the chosen preconditioning voltage. The reason for this advantageous behavior is that hole capture time constants of the electrically active defects become shorter than the used pulse lengths.

5. Conclusions

In summary, we have presented the short-term charge trapping dynamics in SiC MOSFETs of several manufacturers and linked them via the impact on the threshold voltage to the drain-source current variations during pulsed IV-curve measurements. Our results clearly show that device parameters, such as threshold voltage, drain-source current, and on-state resistance, strongly depend on the measurement scheme and precise timing on a microsecond scale. The value of such parameters is influenced by the measurement procedure itself and the short- and long-term history of the gate signal. Without precise timing, the error of a measured on-state resistance is around 10% and without the use of any preconditioning techniques, the error of the threshold voltage can reach several volts.

Furthermore, negative preconditioning is a very effective and tolerant approach for the accelerated removal of trapped electrons with small time constants. Fast recovery of holes trapped during the preconditioning pulse and the insensibility to the preconditioning parameters are major advantages of this method. It has to be emphasized that every measurement following preconditioning is however again strongly dependent on its measurement parameters.

Author Contributions: Conceptualization, T.G. and H.R.; methodology, H.R.; investigation, M.W.F., A.H., K.P. and C.S.; writing—original draft preparation, M.W.F.; writing—review and editing, M.W.F., A.H., K.P., C.S., T.A., H.R. and T.G.; visualization, M.W.F.; supervision, T.G. and W.G.; project administration, H.R. and T.A. All authors have read and agreed to the published version of the manuscript.

Funding: Open Access Funding by TU Wien Bibliothek.

Acknowledgments: The financial support by the Austrian Federal Ministry for Digital and Economic Affairs and the National Foundation for Research, Technology and Development is gratefully acknowledged. Furthermore, the authors acknowledge TU Wien Bibliothek for financial support through its Open Access Funding Programme.

Conflicts of Interest: The authors declare no conflict of interest.

Abbreviations

The following abbreviations are used in this manuscript:

SiC	silicon carbide
Si	silicon
MOSFET	metal-oxide-semiconductor field-effect transistor
BTI	bias temperature instability

Appendix A. Sweeped IV-Curves at High Drain-Source Voltage

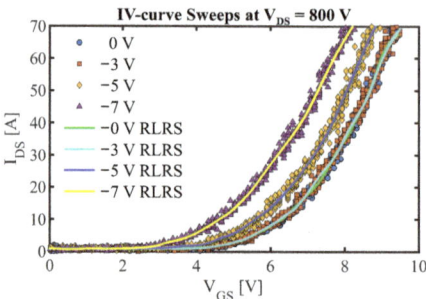

Figure A1. Fast gate voltage sweeps from different initial gate voltages up to 15 V within 5 µs at a drain-source voltage of 800 V at 25 °C and the corresponding smoothed data lines based on robust local regression smoothing (RLRS). While the IV-curves of the initial sweep voltages 0 V and −3 V coincide, the IV-curves corresponding to voltages of −5 V and −7 V are shifted towards lower gate voltages indicating a transient threshold voltage shift.

Appendix B. Base Voltage Dependence of the Drain-Source Current

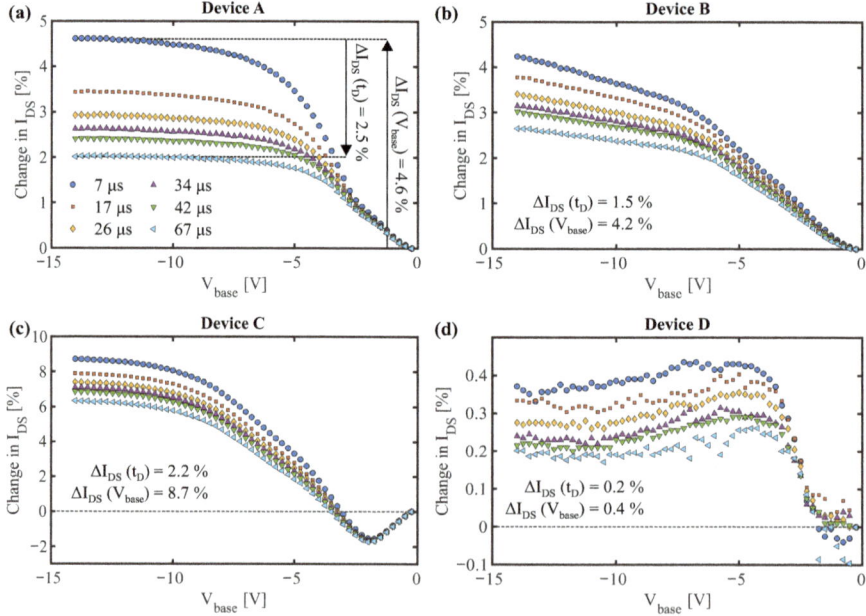

Figure A2. The normalized change of the drain-source current I_{DS} for different delay times t_D of (**a**) device A, (**b**) device B, (**c**) device C, and (**d**) device D. The quantities $\Delta I_{DS}(t_D)$ and $\Delta I_{DS}(V_{base})$ describe the maximum induced change in the drain-source current, whereby the former describes the change due to increasing the delay time of 7 µs and the latter expresses the change due to lowering the base voltage relatively to −0.25 V.

Appendix C. Correlation between Change in Drain-Source Current and the Threshold Voltage Shift

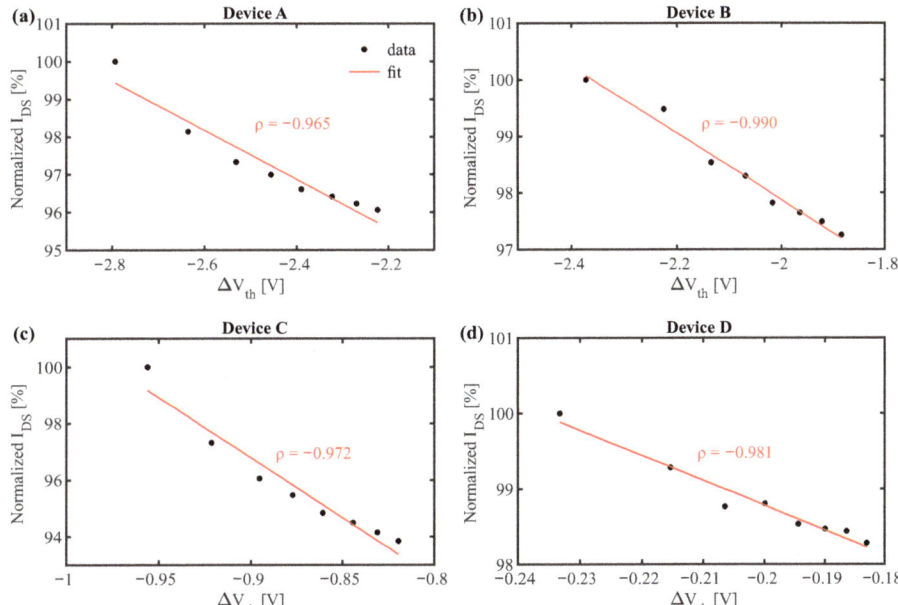

Figure A3. Correlation between the measured change in drain-source current normalized to the first readout and the threshold voltage shift during a 15 V pulse with a base voltage of −12 V. The red lines indicate linear fits. The shown quantity ρ is the Pearson correlation coefficient indicating strong linear correlation for all devices. The subfigures (a)–(d) correspond to the respective devices A–D. For the condition $V_{DS} \ll V_{GS} - V_{th}$, the drain-source current follows $I_{DS} \propto V_{GS} - V_{th}$, implying a linear relationship between the drain-source current and the threshold voltage shift. As the data satisfies both the condition and the linear relationship, it can be concluded that the observed change of the drain-source current indeed originates from the transient threshold voltage shift.

References

1. Afanas'ev, V.V.; Bassler, M.; Pensl, G.; Schulz, M.J.; Stein von Kamienski, E. Band offsets and electronic structure of SiC/SiO$_2$ interfaces. *J. Appl. Phys.* **1996**, *79*, 3108–3114. [CrossRef]
2. Palmour, J.W. Silicon carbide power device development for industrial markets. In Proceedings of the 2014 IEEE International Electron Devices Meeting, San Francisco, CA, USA, 15–17 December 2014; pp. 1.1.1–1.1.8. [CrossRef]
3. Heer, D.; Domes, D.; Peters, D. Switching performance of a 1200 V SiC-Trench-MOSFET in a low-power module. In Proceedings of the PCIM Europe 2016, Nuremberg, Germany, 10–12 May 2016; pp. 53–59.
4. Guo, S.; Zhang, L.; Lei, Y.; Li, X.; Xue, F.; Yu, W.; Huang, A.Q. 3.38 Mhz operation of 1.2kV SiC MOSFET with integrated ultra-fast gate drive. In Proceedings of the 3rd IEEE Workshop on Wide Bandgap Power Devices and Applications, Blacksburg, VA, USA, 2–4 November 2015; Volume 25, pp. 390–395. [CrossRef]
5. Millan, J.; Godignon, P.; Perpina, X.; Perez-Tomas, A.; Rebollo, J. A Survey of Wide Bandgap Power Semiconductor Devices. *IEEE Trans. Power Electron.* **2014**, *29*, 2155–2163. [CrossRef]
6. Peters, D.; Siemieniec, R.; Aichinger, T.; Basler, T.; Esteve, R.; Bergner, W.; Kueck, D. Performance and Ruggedness of 1200 V SiC-Trench-MOSFET. In Proceedings of the 2017 29th International Symposium on Power Semiconductor Devices and IC's, Sapporo, Japan, 28 May–1 June 2017; pp. 239–242. [CrossRef]
7. Yano, H.; Nakao, H.; Hatayama, T.; Uraoka, Y.; Fuyuki, T. Increased Channel Mobility in 4H-SiC UMOSFETs Using On-Axis Substrates. *Mater. Sci. Forum* **2007**, *556–557*, 807–810. [CrossRef]

8. Anwar, S.; Wang, Z.J.; Chinthavali, M. Characterization and Comparison of Trench and Planar Silicon Carbide (SiC) MOSFET at Different Temperatures. In Proceedings of the 2018 IEEE Transportation Electrification Conference and Expo (ITEC), Long Beach, CA, USA, 13–15 June 2018; pp. 1039–1045. [CrossRef]
9. Aichinger, T.; Schmidt, M. Gate-oxide reliability and failure-rate reduction of industrial SiC MOSFETs. In Proceedings of the 2020 IEEE International Reliability Physics Symposium (IRPS), Dallas, TX, USA, 28 April–30 May 2020; pp. 1–6. [CrossRef]
10. Puschkarsky, K.; Grasser, T.; Aichinger, T.; Gustin, W.; Reisinger, H. Review on SiC MOSFETs High-Voltage Device Reliability Focusing on Threshold Voltage Instability. *IEEE Trans. Electron. Devices* **2019**, *66*, 4604–4616. [CrossRef]
11. Schleich, C.; Berens, J.; Rzepa, G.; Pobegen, G.; Rescher, G.; Tyaginov, S.; Grasser, T.; Waltl, M. Physical Modeling of Bias Temperature Instabilities in SiC MOSFETs. In Proceedings of the 2019 IEEE International Electron Devices Meeting (IEDM), San Francisco, CA, USA, 7–11 December 2019; pp. 486–489. [CrossRef]
12. Rescher, G.; Pobegen, G.; Aichinger, T.; Grasser, T. On the subthreshold drain current sweep hysteresis of 4H-SiC nMOSFETs. In Proceedings of the 2016 IEEE International Electron Devices Meeting (IEDM), San Francisco, CA, USA, 3–7 December 2016; pp. 10.8.1–10.8.4. [CrossRef]
13. Puschkarsky, K.; Grasser, T.; Aichinger, T.; Gustin, W.; Reisinger, H. Understanding and Modeling Transient Threshold Voltage Instabilities in SiC MOSFETs. In Proceedings of the 2018 IEEE International Reliability Physics Symposium (IRPS), Burlingame, CA, USA, 11–15 March 2018; pp. 3B.5-1–3B.5-10. [CrossRef]
14. Puschkarsky, K.; Reisinger, H.; Aichinger, T.; Gustin, W.; Grasser, T. Threshold voltage hysteresis in SiC MOSFETs and its impact on circuit operation. In Proceedings of the 2017 IEEE International Integrated Reliability Workshop (IIRW), Fallen Leaf Lake, CA, USA, 8–12 October 2017. [CrossRef]
15. Lelis, A.J.; Green, R.; Habersat, D.B. SiC MOSFET threshold-stability issues. *Mater. Sci. Semicond. Process.* **2018**, *78*, 32–37. [CrossRef]
16. Aichinger, T.; Rescher, G.; Pobegen, G. Threshold voltage peculiarities and bias temperature instabilities of SiC MOSFETs. *Microelectron. Reliab.* **2018**, *80*, 68–78. [CrossRef]
17. Rescher, G.; Pobegen, G.; Aichinger, T.; Grasser, T. Preconditioned BTI on 4H-SiC: Proposal for a Nearly Delay Time-Independent Measurement Technique. *IEEE Trans. Electron Devices* **2018**, *65*, 1419–1426. [CrossRef]
18. Reisinger, H.; Blank, O.; Heinrigs, W.; Mühlhoff, A.; Gustin, W.; Schlünder, C. Analysis of NBTI Degradation- and Recovery-Behavior Based on Ultra Fast VT-Measurements. In Proceedings of the 2006 IEEE International Reliability Physics Symposium Proceedings, San Jose, CA, USA, 26–30 March 2006; pp. 448–453. [CrossRef]
19. Basler, T.; Heer, D.; Peters, D.; Aichinger, T.; Schörner, R. Practical Aspects and Body Diode Robustness of a 1200 V SiC Trench MOSFET. In Proceedings of the PCIM Europe 2018, International Exhibition and Conference for Power Electronics, Intelligent Motion, Renewable Energy and Energy Management, VDE, Nuremberg, Germany, 5–7 June 2018; pp. 536–542.
20. Gruber, G.; Cottom, J.; Meszaros, R.; Koch, M.; Pobegen, G.; Aichinger, T.; Peters, D.; Hadley, P. Electrically detected magnetic resonance of carbon dangling bonds at the Si-face 4H-SiC/SiO$_2$ interface. *J. Appl. Phys.* **2018**, *123*, 161514. [CrossRef]

Publisher's Note: MDPI stays neutral with regard to jurisdictional claims in published maps and institutional affiliations.

© 2020 by the authors. Licensee MDPI, Basel, Switzerland. This article is an open access article distributed under the terms and conditions of the Creative Commons Attribution (CC BY) license (http://creativecommons.org/licenses/by/4.0/).

Review

Numerical Simulation of Ammonothermal Crystal Growth of GaN—Current State, Challenges, and Prospects

Saskia Schimmel [1,2,*], Daisuke Tomida [1], Tohru Ishiguro [3], Yoshio Honda [1], Shigefusa Chichibu [1,3] and Hiroshi Amano [1]

1. Institute of Materials and Systems for Sustainability, Nagoya University, Nagoya 464-8601, Japan; tomida@imass.nagoya-u.ac.jp (D.T.); honda@imass.nagoya-u.jp (Y.H.); chichibu@tohoku.ac.jp (S.C.); amano@nuee.nagoya-u.ac.jp (H.A.)
2. International Research Fellow of Japan Society for the Promotion of Science, Nagoya University, Nagoya 464-8601, Japan
3. Institute of Multidisciplinary Research for Advanced Materials, Tohoku University, Sendai 980-8577, Japan; toru.ishiguro.b5@tohoku.ac.jp
* Correspondence: sas.schimmel@imass.nagoya-u.ac.jp

Citation: Schimmel, S.; Tomida, D.; Ishiguro, T.; Honda, Y.; Chichibu, S.; Amano, H. Numerical Simulation of Ammonothermal Crystal Growth of GaN—Current State, Challenges, and Prospects. *Crystals* **2021**, *11*, 356. https://doi.org/10.3390/cryst11040356

Academic Editor: Michael Waltl

Received: 8 March 2021
Accepted: 24 March 2021
Published: 30 March 2021

Publisher's Note: MDPI stays neutral with regard to jurisdictional claims in published maps and institutional affiliations.

Copyright: © 2021 by the authors. Licensee MDPI, Basel, Switzerland. This article is an open access article distributed under the terms and conditions of the Creative Commons Attribution (CC BY) license (https://creativecommons.org/licenses/by/4.0/).

Abstract: Numerical simulations are a valuable tool for the design and optimization of crystal growth processes because experimental investigations are expensive and access to internal parameters is limited. These technical limitations are particularly large for ammonothermal growth of bulk GaN, an important semiconductor material. This review presents an overview of the literature on simulations targeting ammonothermal growth of GaN. Approaches for validation are also reviewed, and an overview of available methods and data is given. Fluid flow is likely in the transitional range between laminar and turbulent; however, the time-averaged flow patterns likely tend to be stable. Thermal boundary conditions both in experimental and numerical research deserve more detailed evaluation, especially when designing numerical or physical models of the ammonothermal growth system. A key source of uncertainty for calculations is fluid properties under the specific conditions. This originates from their importance not only in numerical simulations but also in designing similar physical model systems and in guiding the selection of the flow model. Due to the various sources of uncertainty, a closer integration of numerical modeling, physical modeling, and the use of measurements under ammonothermal process conditions appear to be necessary for developing numerical models of defined accuracy.

Keywords: ammonothermal; crystal growth; numerical simulation; gallium nitride; computational fluid dynamics; conjugated heat transfer; natural convection; buoyancy; solvothermal; hydrothermal

1. Introduction

The ammonothermal method has initially been developed as a tool for the synthesis and recrystallization of metal amides and metal nitrides, taking advantage of enhanced solubilities of inorganic substances in supercritical ammonia containing mineralizers [1]. Starting from 1995 [2], the ammonothermal process has been increasingly researched as a method for the growth of GaN bulk crystals [3–8]. Depending on the choice of mineralizer, it is possible to obtain GaN in its cubic or wurtzite structure [9], although most research has focused on wurtzite GaN. The ammonothermal method is recognized as particularly promising for the growth of GaN with high structural quality by using near-equilibrium conditions [6,10]. On the contrary, impurity concentration and thus conductivity control are particularly challenging due to the closed growth system [11]. GaN has already found widespread commercial application in blue light-emitting diodes, though LEDs are usually grown on foreign substrates [12]. Development of growth technologies for bulk GaN is driven by those device applications for which high dislocation density in the heteroepitaxial nitride structures is a critical issue, such as laser diodes and power electronic devices (in particular, vertical high-power transistors and diodes) [13–16]. Like the very successful hydrothermal method for the growth of oxides such as quartz and ZnO [1,17], the ammonothermal method also has the potential to grow a large number of crystals simultaneously [17]. Due to its scalability, the ammonothermal method is still believed to have the potential to become a strong competitor for halide vapor phase epitaxy (HVPE), which is the most common method for bulk GaN growth at present [13]. Recent progress in ammonothermal GaN growth includes a demonstration of scalability to pilot production for simultaneous growth of likely about 100 boules in one reactor [6], a masking technique for circumventing issues related to growth on different facets [18], growth of nearly bow-free crystals (radius of curvature: 1460 m) at pressures as low as 100 to 120 MPa [19], and growth of nearly 4-inch size crystals while keeping off-angle distributions as small as ±0.006° along both a-axis and m-axis [20]. Besides its use for the growth of bulk GaN, the ammonothermal technique is increasingly being utilized for exploratory syntheses of various binary, ternary, and multinary nitride and oxynitride materials [21–23], including nitride semiconductors composed of earth-abundant elements [24].

A technical challenge for understanding and optimizing ammonothermal syntheses lies in the difficulty of experimental access to the interior of the autoclave during the process, which is due to the experimental parameters of about 600 °C and 100 to 400 MPa. Therefore, numerical modeling is an important tool for clarifying the actual experimental conditions such as temperature distribution and flow field inside the autoclave. A number of research groups have conducted numerical studies of temperature and flow field [25–28], partially including further aspects such as the concentration of metastable intermediates [29] and growth rates [29,30]. However, critical issues have not been studied thoroughly in the literature, and validation has been applied only to an extremely limited extent. Therefore, the accuracy of such numerical results remains unclear, preventing simulations from delivering their full potential impact on further development and use of the ammonothermal method for bulk crystal growth. Moreover, it should be noted that natural convection in cylindrical enclosures with laterally heated walls in general is not well studied to date [28,31].

The ammonothermal growth conditions and growth process represent a complex multiphysics problem. For a glossary of common terms and information on multiphysics simulations in general, the reader is referred to a respective review [32]. Moreover, a recent review on conjugate heat transfer simulations can be found in [33], and solvers for coupled porous media flow have recently been reviewed in [34].

Besides the difficult experimental access to internal conditions, the limited number of experiments also represents a bottleneck for scientific and technological progress. Costs for growth experiments are driven by investment costs for corrosion-resistant high-pressure equipment and costs for consumables including seeds and gaskets. A breakthrough in the trustworthiness of numerical simulations of growth conditions and ideally also the growth process itself could therefore tremendously speed up further development.

In the second section of this review, the general functionality of ammonothermal bulk growth of GaN will be described. The third section gives an overview and discussion of numerical studies of flow and temperature field in the literature. Simulations of the growth process or subprocesses thereof beyond the thermal and flow field are reviewed in Section 4. The fifths section covers approaches for validation of ammonothermal simulations. The sixth section shines some light on further, potentially relevant aspects that may affect the accuracy of current CFD (Computational Fluid Dynamics) simulations of ammonothermal growth systems. Finally, a brief conclusion and outlook is be given.

2. Functionality of the Ammonothermal GaN Growth Process

If employed for bulk crystal growth, the ammonothermal method is based on the use of buoyancy as the driving force for convective mass transport and a solubility difference as the driving force for recrystallization (i.e., dissolution of the nutrient and crystallization on the seed crystals). Figure 1 schematically illustrates the functionality of ammonothermal growth in the current common understanding, for growth in the temperature range of retrograde solubility. It should be noted that the transport of gallium requires the assistance of a mineralizer to promote its solubility, likely by the formation of complex ions [35–38], which appear to aggregate under ammonothermal process conditions [38,39]. The temperature dependence of solubility depends on the mineralizer and may comprise both a range of normal solubility at lower temperatures and retrograde solubility at higher temperatures. Such transitions from normal to retrograde solubility have been observed experimentally at least when using the ammonium halides NH_4Cl [40] or NH_4F [4]. Besides high-temperature versions of ammonoacidic growth, ammonobasic growth also operates in the retrograde solubility range [9], albeit at lower absolute temperatures and higher pressures than ammonoacidic growth.

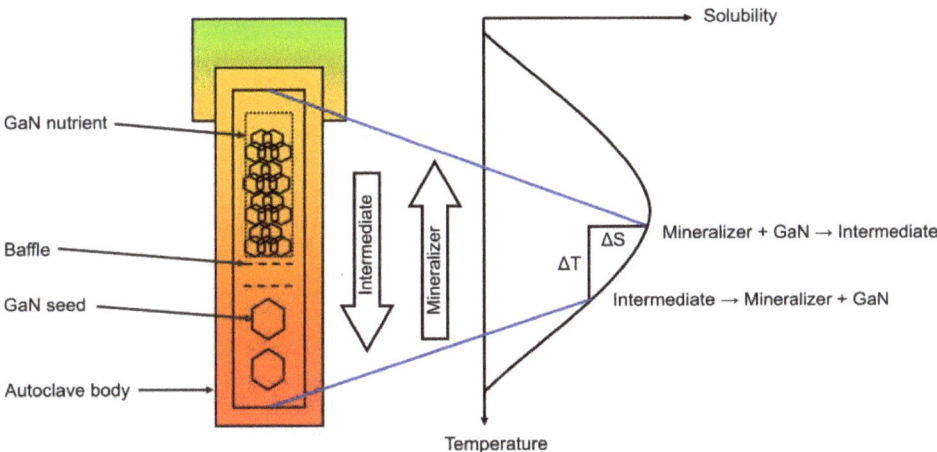

Figure 1. Basic functionality of ammonothermal GaN bulk growth process. Recrystallization of GaN takes place in an autoclave using a temperature gradient to create a driving force for dissolution and crystallization in different zones of the autoclave. Mass transport of Ga is achieved by the formation of soluble intermediates and their transport by natural convection of the fluid.

Although numerical simulations have so far focused on the (quasi-)steady stage of the growth process, one should be aware that this is not the only relevant stage of an ammonothermal growth run. Experimental procedures suggest that etch-back of the seeds during early stages of the experiment is also important, given that experimenters deliberately use a back-etching step prior to allowing nucleation on the seeds [41]. Therefore, transient stages of the experiment such as ramp-up and sometimes a process step with inverted external temperature gradient appear to be critical steps of the process. A schematic representation of a temperature program suitable for growth in the retrograde solubility range is depicted in Figure 2, following the temperature program presented by Grabianska et al. for basic ammonothermal crystallization developed at Institute of High Pressure Physics Polish Academy of Sciences [41]. Besides giving an example of the externally applied temperature program, Figure 2 also illustrates the fact that internal temperatures differ from externally controlled temperatures. The schematically indicated mean internal temperatures are based on the assumption that convective heat transfer will reduce internal thermal gradients compared to the externally applied ones, which is in agreement with a numerical study by Chen et al. [42], among others. The inset in Figure 2 visualizes an effect that is not only likely relevant to the growth process itself but also of practical importance for numerical simulations. Several numerical studies suggest that the fluid flow can be highly oscillatory, at least in certain regions of the autoclave, such as in the vicinity of the baffle. This is in agreement with experimental observations, though the experimental results suggest that the temperature and flow fluctuations may not follow a regular, oscillatory pattern [43]. From a growth point of view, instabilities of fluid flow may have different effects depending on their location. Mirzaee et al. consider them to be potentially beneficial for mass transport across the baffle if occurring in that region but assume them to be detrimental if occurring in the vicinity of the seeds [29]. In the authors' opinion, instability of fluid flow should not be required even in the baffle area if the areas with upward and downward flows are well-balanced and allow for sufficient mass flow; however, it does not seem fully clarified whether they are avoidable under all other constraints. From the viewpoint of numerical simulations, the fluid flow's susceptibility to instability poses a challenge because it makes transient simulations prone to requiring rather small timesteps while instable flow occurs. The need for small timesteps, in turn, makes it difficult to keep computation times of transient process simulations in a reasonable order of magnitude.

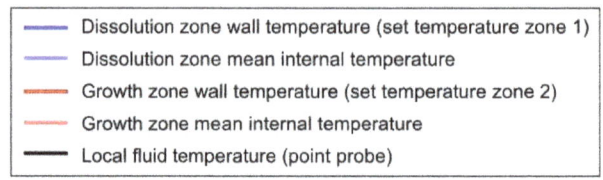

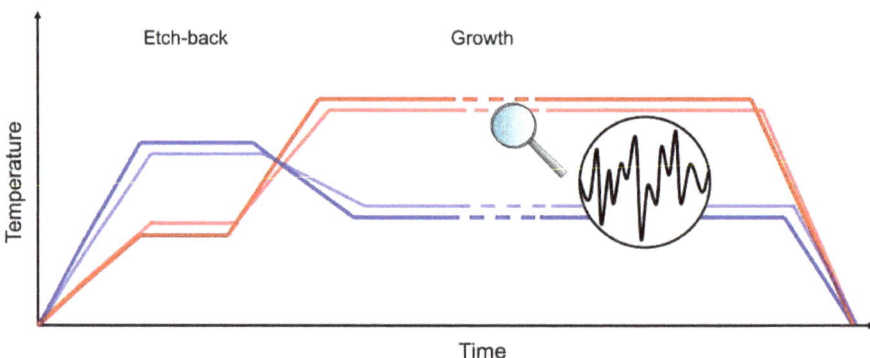

Figure 2. Schematic representation of the temperature program used for ammonothermal growth of GaN in the retrograde solubility range. The externally controlled temperature program (set temperatures) is based on [41]. The mean internal temperatures are not known but intended to indicate that there is a difference between internal and external temperatures, with the internal gradient likely being smaller than the externally applied gradient due to convective heat exchange between the two zones. The round inset visualizes that local fluid temperatures are often unstable and may not even follow a strictly oscillatory pattern, on the basis of [43] and references therein.

3. Simulations of Fluid Flow and Temperature Field

In the following, literature on simulations of fluid flow and temperature field will be reviewed. A description of ammonothermal equipment will also be included at the beginning, in order to explain and discuss common simplifications in simulations.

3.1. Simulation Domain and Geometry

The experimental geometry of ammonothermal setups for bulk growth of GaN is generally some variation of the setup depicted in Figure 3a, which is based on several articles from the literature [4,20,44,45] and the experience of the authors. The internal configuration shown in the figure is valid for the case of retrograde solubility of GaN. Essential elements of an ammonothermal growth setup comprise the autoclave in the vertical orientation, a furnace that allows a temperature gradient to be established in the axial direction of the autoclave, and a head assembly with peripheral devices. The latter are needed for pressure monitoring and the removal and introduction of gaseous substances and are usually connected by stainless steel pipes.

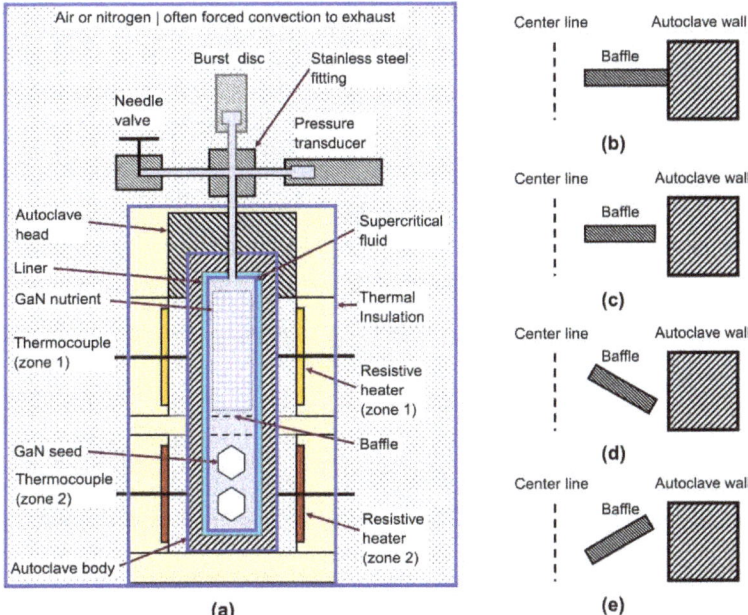

Figure 3. Setup for bulk crystal growth of GaN using an ammonothermal method. (a) Complete setup consisting of autoclave with head assembly in a furnace, which are usually surrounded by some enclosure filled either with ambient air or nitrogen. The illustration assumes retrograde temperature dependence of solubility (in case of normal solubility, the locations of GaN seeds and nutrient would be inverted). Thick blue lines represent possible choices for the boundaries of the simulation domain. Parts drawn with gray line color (burst disc, insulation of autoclave head, liner) are optional and may not be used, depending on the mineralizer and safety concept. (b–e) Baffle geometries considered in simulation literature: (b) ring-shaped baffle directly attached to the autoclave wall, (c) ring-shaped baffle with a gap between baffle and autoclave wall, (d) funnel-shaped baffle with a gap between baffle and autoclave wall.

The entire setup is commonly run inside an enclosure (or adequate separate room free of personnel) in order to protect operators from shrapnel and toxic gases in case of mechanical failure of high-pressure parts. Such enclosures may be run as a fume-hood-like system with ambient air if permittable from a safety point of view. Since common safety concerns include prevention of formation of explosive gas mixtures in the proximity of possible sources for ignition and leakage of toxic gases outside the enclosure, high air turnovers are usually chosen if using air-filled enclosures [46]. Alternatively, enclosures can be designed to provide an oxygen-free (e.g., nitrogen) atmosphere, e.g., for autoclaves made of molybdenum alloys [47].

Besides the reaction medium, the interior of the autoclave contains GaN nutrient, GaN seeds, and (ideally inert) parts for flow control and for positioning GaN nutrient and seeds. Flow control is achieved using one (e.g., [48]) or more (e.g., [44,45]) baffles, which are often axially symmetric. However, examples of non-axially symmetric baffle assemblies are occasionally shown [45]. Numerical simulations have focused on axially symmetric baffles [25–27,29,30,42,49–53], in part likely for taking advantage of axial symmetry in 2D simulations. Different types of baffle configurations that have been applied in the simulation literature of ammonothermal GaN growth are depicted in Figure 3b–e. Specifically, experimental setups typically exhibit a thin gap in between the inner autoclave wall and the outer edge of the baffle (Figure 3c). Some numerical studies have neglected this gap (likely for simplifying the mesh and avoiding related issues). However, others have included it and found significant flow to occur through this gap [27,52,54], which is in accordace with an experimental observation by Masuda et al. that a small gap between baffle and autoclave wall is required for successful crystal growth experiments [53]. Moreover, funnel-shaped baffles (Figure 3d–e) have been studied numerically in 2D [51] and 3D [26] but there seem to be no experimental studies on inclined baffle geometries.

For numerical simulations, different sub-regions of the experimental setup can in principle be selected as the simulation domain, depending on the questions to be answered as well as on computational resources. Domain boundary choices found in the literature are the inner or outer autoclave wall, the wall of the furnace or a gas volume surrounding the furnace and head assembly. The implications of these choices as well as examples are discussed in Section 3.3, as the definition of domain boundary is closely linked to the definition of domain boundary conditions.

Components that have so far usually been neglected in simulation literature comprise the liner or corrosion-resistant capsule (if applicable), as well as the head assembly. However, Moldovan studied an ammonothermal setup with a silver liner [55]. Moreover, simplifications of geometry are applied, especially for obtaining axial symmetry. Typical examples are the head assembly and pipes, which are completely neglected in most models, as well as the seed crystals. The effects of the head assembly, and especially its horizontal parts with the biggest lack of axial symmetry, can be expected to have rather symmetric effects on the thermal field inside the autoclave. In the author's opinion, the most relevant effect of the head assembly should be thermal losses, because the parts are not insulated and kept at temperatures from room temperature to about 150 °C due to the limited high-temperature strength of stainless steels. Changes in the magnitude of heat losses through the head assembly have been found to trigger changes in the crystallization of reaction products inside the nickel base alloy tube that connects the autoclave head and the central fitting of the head assembly [56]. The magnitude of such heat losses must be expected, depending on the temperature of the ambient in its surrounding, the gas turnover in the vicinity of the furnace, and on the presence of heat-conducting connections to cooler parts (such as for filling or for fixing the setup mechanically, depending on the specific facilities). Those heat losses are likely approximately axial symmetric. However, they are likely not negligible, and thus, sound estimates of their magnitude may be necessary for obtaining accurate results.

3.2. Axisymmetric 2D Calculations versus 3D Calculations

It should be noted that there is still an active discussion on the need for 3D calculations and the limitations of 2D calculations for simulations of thermal field and flow field in ammonothermal systems. Several groups have so far focused on 2D calculations [25,27,29,57], taking advantage of their much lower computational cost. In their 2008 review on modeling of crystal growth processes in general, Chen et al. suggest the coupling of a 2D axisymmetric global thermal field model with a 3D model for the fluid flow, among others, for ammonothermal growth modeling [58]. Experimentally validated simulations of an ambient pressure model of a hydrothermal crystal growth reactor reveal transient 3-dimensional fluid flow but also conclude that the time-averaged flow and temperature fields are axially symmetric, if only axially symmetric solids are considered [59]. However, it appears reasonable to expect a need for 3D calculations at least if effects of not axially symmetric parts are considered. This is the case for seeds, which are usually flat cuboids initially and tend to develop facets during the growth run [5]. Very recently, there have been reports on the use of round seeds and limiting the growth to one direction in order to eliminate the stress generated in crystals due to inhomogeneous incorporation of impurities [41]. In this case, however, it appears that the axis of the round seeds remains perpendicular to the axis of the autoclave and thus the symmetry axis of the simulation domain. Thus, the seeds remain a not axially symmetric element. One of the objects with the largest deviations from axial symmetry therefore also constitutes a primary region of interest to the crystal grower. A few groups have done 3D calculations and concluded that 3D calculations are necessary for obtaining accurate results on the heat and mass transport inside the autoclave [26,31]. It should be noted that these simulations did not even contain any not axially symmetric solids [26,31]. The study by Enayati et al. uses an experimental model with water as both the working fluid inside the enclosure and as a means of wall temperature control [31]. Using this experimental transparent setup, Enayati et al. present validation data obtained using tracer particles and a laser-based imaging system [31]. The validation data fit the experimental data rather well, supporting the numerical results as well as the chosen heater-long constant temperatures as a boundary condition. It should, however, be noted that the adiabatic walls at the bottom and especially at the top of the enclosure may represent a major deviation from actual ammonothermal growth setups. As recently reported by Schimmel et al. thermal losses through the autoclave head and head assembly appear to have a significant influence on the temperature distribution of the autoclave walls [57]. Therefore, both the experimental and the numerical model may not be an accurate representation of a typical solvothermal growth reactor. Despite the decent validation, it is therefore not clear to what extent the results can be applied to ammonothermal growth. The study by Erlekampf et al. provides validation data by several internal temperature measurements in an actual ammonothermal autoclave [26]; however, the temperature configuration used (inverted temperature gradient in relation to common practice in solvothermal bulk crystal growth) raises the question of whether the agreement between numerical and experimental data remains similar if a buoyancy-promoting orientation of the temperature gradient was used.

3.3. Boundary Conditions

Depending on the choice of simulation domain, boundary conditions can be fixed temperatures at the domain boundaries or other objects, heat sources of fixed or adjustable power density, and conditions pertaining to gas exchange and radiative heat losses across the domain boundaries. A schematic overview of types of boundary condition found in the literature is given in Figure 4.

Until recently, almost all simulation studies in the literature chose either the inner or the outer autoclave wall as the boundary of the simulation domain [25,27,29,60]. Most studies employ fixed temperatures for major sections of the outer autoclave walls [25,29,50], usually corresponding to the length of heaters. A study of hydrothermal fluid flow uses a step function to account for thermal gradients in autoclave walls but fixes temperature at the inner wall of the autoclave [61], therefore completely neglecting the influence of convective heat transfer on inner wall temperatures. A validated simulation of a hydrothermal model system has been reported by Ursu et al. and works with fixed temperatures for hot and cold zone [62]. However, the model setup used by Ursu et al. can be expected to create rather homogeneous wall temperatures because a recirculating water bath is used for heating. Thus, the model setup may be an accurate representation of the numerical model but not necessarily of the actual hydrothermal growth setup. Erlekampf et al. were the first to publish a numerical study of an ammonthermal reactor that includes the furnace in a 2D simulation of the global thermal field and uses heater power density as the heat source [26]. Moreover, they provided at least some validation using the actual ammonothermal setup [26]. However, convective heat transfer was apparently neglected in the 2D simulation. For their subsequent 3D simulation, Erlekampf et al. chose the inner autoclave wall as the domain boundary, using the temperature distribution from their 2D model as the boundary condition [26]. Most recently, Schimmel et al. reported a first 2D simulation that expands the domain boundary even further in order to investigate the temperature distribution in the autoclave walls [57].

In summary, the different groups made different choices about what to neglect in order to create a sufficiently simple model. However, there is no comprehensive information in the literature as to which choices are most reasonable and how they affect the results. Based on the recent publication by Schimmel et al. [57], it appears that outer wall temperatures can be an effective and reasonably efficient choice but require knowledge of the temperature distribution in the outer autoclave walls rather than just knowledge of the heater set temperatures.

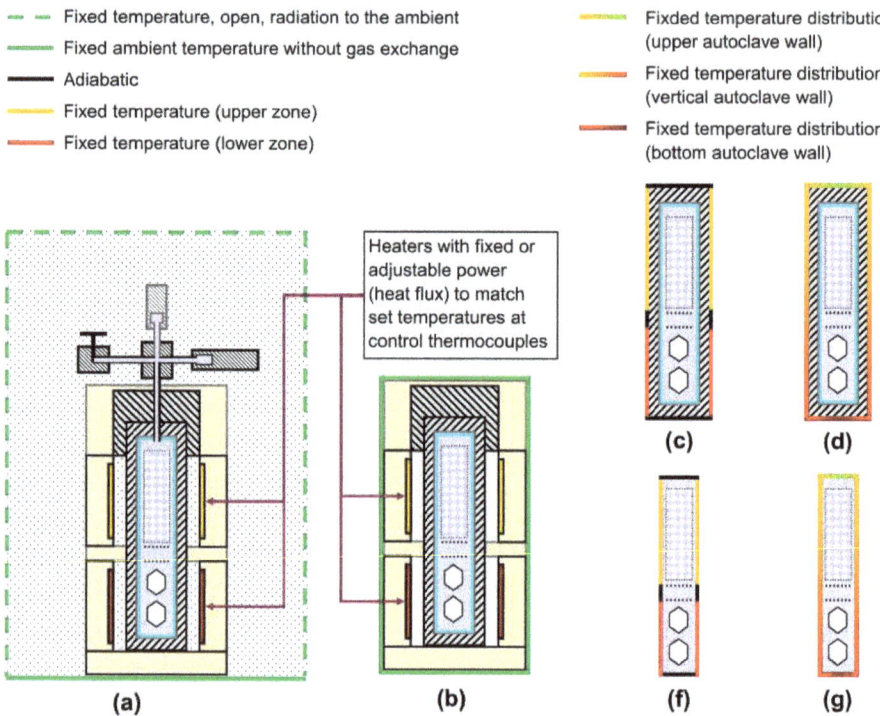

Figure 4. Boundary conditions for different choices of the simulation domain. (**a,b**): fixed heater power densities; (**a**) allows for gas exchange via open boundaries as well as radiative losses to the ambient, whereas (**b**) uses a fixed temperature of the domain boundary, (**c,f**) heater-long constant fixed temperatures and adiabatic walls elsewhere, (**d,g**), fixed temperature distributions at all walls, and (**c,d**) show simulation domain including the autoclave wall, whereas (**f,g**) represent simulation domain excluding the autoclave walls.

3.4. Physics and Models Thereof

In principle, the physics involved that may influence the thermal field and flow fields include heat conduction in solids and fluids, natural convection, and potentially radiative heat transfer. The relative importance of those effects depends on the choice of simulation domain. Specifically, radiation is relevant if the furnace is included in the simulation and if the setup features air-filled gaps in between heat sources and autoclave walls as in [57]. In most published studies, however, the border of the simulation domain was placed at the outer or inner autoclave walls, and radiation was not deemed to be relevant.

Regarding natural convection inside the autoclave, appropriate models for the fluid flow must be chosen for the solid-free regions as well as for the porous medium constituted by the nutrient. In principle, the Navier-Stokes equations need to be solved. As recently discussed in further detail in [57], fluid flow under ammonothermal process conditions is likely in the transition range between laminar and turbulent flow, and turbulent flow is very likely to occur at least in some regions of the cavity or under some experimental conditions. In practice, with limited computational resources, it is important to apply suitable approximations rather than Direct Numerical Simulation (DNS), which simulates the whole range of the turbulent statistical fluctuations at all relevant physical scales [63]. Approximations for turbulent flow include Large Eddy Simulation (LES) and Reynolds Averaged Navier-Stokes (RANS) models ([63], p. 90). The former, LES, directly simulates turbulent fluctuations above a certain length scale and models those below this length scale ([63], p. 87). The modeling at the so-called subgrid scale employs semi-empirical laws ([63], p. 87). The latter, RANS, is restricted to the simulation of time-averaged turbulent flow ([63], p. 90). While being computationally most efficient, it should be noted that RANS simulations require empirical or at least semi-empirical information on the turbulence structure and its relation to the averaged flow ([63], p. 88f). A review of RANS models can be found in [64]. Regarding simulations of solvothermal flow fields, Enayati et al. used 3D LES, among others, for studying the effects of baffle on temperature distribution and flow field [31,49] as well as for studying the effects of a rack and seeds [65] and of reactor size [60]. For comparing the results of 2D and 3D calculations, Enayati et al. employed a RANS model [31]. Likewise, the majority of ammonothermal flow and temperature field calculations have utilized different RANS models. Pendurti et al. estimated the flow to be turbulent and applied a renormalized group k-ε model [30]. Masuda et al. use a modified production k-ε model to model turbulent flow in ammonothermal and hydrothermal growth, which compensates for the over estimation of the formation near the stagnation point that appears in the standard k-ε equations [51,66]. Enayati et al. apply k-ω shear stress transport (SST) turbulent model for studying the effects of position and permeability of a porous medium in a laterally heated reactor for crystal growth [67]. Recently, Schimmel et al. have applied the relatively simple, yet computationally efficient LVEL (Length-VELocity) turbulence model for studying global heat transfer in an ammonothermal growth setup including the furnace [57]. The LVEL model constitutes a zero-equation low Reynolds number turbulence model, which is valid over the laminar, transitional, and turbulent flow regimes and is particularly well-suited for fluid domains cluttered with solids [68], thus facilitating the study of the complete ammonothermal growth setup in a computationally affordable way. Viewing these studies, it is, however, important to keep in mind that there is a lack of knowledge on fluid properties under process conditions, which causes uncertainties in the dimensionless numbers characterizing the flow. Given the lack of experimentally validated flow simulations in conjunction with the uncertainties in fluid properties, it remains uncertain what flow models should be applied, and whether those remain the same for all experimental conditions. Quantification of uncertainties due to various modeling assumptions (including but not limited to turbulence models) is also lacking. For RANS models in general CFD applications, however, a review of the quantification of model uncertainty has recently been published by Xiao et al. [69].

Besides models for free laminar or turbulent flow, porous media flow occurs within the GaN nutrient and is coupled with the free flow through all other fluid-filled regions of the cavity. Grain sizes of nutrients in experimental research are rarely disclosed; however, Pimputkar et al. state 1.0 to 2.8 mm nutrient grain sizes for a 25 mm by 280 mm cavity, for ammonobasic growth using sodium as the mineralizer [70]. The scarcity of disclosure by experimenters suggests that nutrient grain size is a critical variable. This appears perfectly reasonable, knowing that the flow in the nutrient depends on the product of the Grashof and Darcy numbers, which is proportional to the square of the average diameter of particles [71]. The flow strength in the porous layer can be enhanced by increasing the size of the particles, or by putting particles in bundles as in the hydrothermal growth [71]. Indications of applying such a bundling approach appear occasionally in figures in publications on experimental results (e.g., Figure 5a in [41]), but no details or underlying considerations are commonly given. In numerical simulations, Masuda et al. considered a porous medium with a grain diameter of 5 mm and a (nondimensional) porosity of 0.7 [53]. In their model, the drag that is produced by the fluid passing through the porous medium is considered as pressure drop, which is determined by the equation of Ergun [53]. Darcy's equation with Brinkman extension to account for viscous diffusion [72] (relevant in the vicinity of interfaces [73]) and Forchheimer extension for inertial effects at the microscale [73,74] was applied by Mirzaee et al. [29], Chen et al. [52] and Enayati et al. [67]. Erlekampf et al. used the Darcy model with only the Forchheimer extension [26]. Both Forchheimer and Ergun equations include both viscous and inertial effects that can describe porous media flow in low- to high-velocity regimes for laminar, transitional, and turbulent flow [75]. Deviations appear to occur if the flow has a pronounced contracting/expanding character, which is better captured by the Forchheimer equation [76]. A contracting/expanding character of the flow is promoted by high localization of the pressure drop in the gap between adjacent particles and thus is most pronounced at low porosities close to maximum packing [76]. Therefore, using Ergun or Darcy's law with Forchheimer extension appears equally justified if the porosity of the nutrient is not exceptionally low. As for Brinkman's extension, Auriault showed in a focused analysis that the domain of validity is limited to flows through swarms of fixed particles or fixed beds of fibers at very low concentration and under further specific conditions [77]. Similarly, Battiato et al. conclude that the validity domain of Brinkman's equation corresponds to porous media with very large porosity and very small solid concentration [74]. Thus, there appears to be no benefit of using Brinkman extension unless the porous medium has or develops a large porosity (which may, however, occur at late stages of a growth run).

3.5. Discretization in Space and Time

Discretization in space is usually done by the finite volume method (FVM). This method discretizes the integral formulation of the conservation laws directly in physical space and uses cell-averaged values as its main numerical quantities ([63], p. 203), with the unknowns defined either at the cell centers or at the mesh nodes (termed cell-centered and vertex-centered, respectively) [78]. An alternative approach would be the use of the finite element method (FEM), which uses the local function values at mesh points as its main numerical quantities ([63], p. 203). The reasons for the popularity of the FVM in CFD are its generality, its conceptual simplicity, and its ease of implementation for arbitrary structured or unstructured grids ([63], p. 203). Regarding the computational costs, there are relatively few studies, with contradictory results. Gohil et al. who compared FVM and FEM calculations of a complex geometry using an unstructured mesh and considering both steady and oscillatory flows, state that their FEM calculation arrived at the pre-specified target accuracy much faster than its FVM counterpart [79]. Gohil et al. conclude that the faster convergence of FEM is likely in part due to the coupled treatment of the mass and momentum equations, which is a fundamental property of FEM [79]. Others have reported opposite observations [80,81]. However, the computation time to reach a specified accuracy is rarely reported, even though it appears to be the most reasonable metric for such a comparison.

While structured grids offer advantages in the efficient use of memory and potentially also time, structured grids have drawbacks when complex geometries are to be meshed [82]. In the case of the ammonothermal method, this becomes increasingly relevant as one attempts a more realistic implementation of internal solids (such as several seeds, seeds in 3D, baffles that are not necessarily always axially symmetric, etc.). Moreover, structured grids cannot grade cell size as rapidly as unstructured grids, causing an increased number of cells [82] if there are regions that require very different cell sizes. In the ammonothermal case, this can, for instance, be relevant to global heat transfer simulations because of the different length scales inside the autoclave, in the furnace and outside the furnace.

For the interior of the autoclave, some estimates of functional cell sizes can be extracted from the literature, though they can of course vary considerably depending on the model and question to be answered. Pendurti et al. state that a mesh size of 0.4 mm is sufficient to resolve the laminar sublayer within the turbulent boundary layer in the proximity of a seed crystal [30]. Mirzaee et al. who employed a quadrilateral mesh, chose a mesh size of 50×250 in radial and vertical directions on the basis of a mesh convergence study using GaN growth rate as the monitored quantity for determining mesh convergence [29,83]. This corresponds to average cell dimensions of 0.762 mm in the radial and 1.524 mm in vertical directions, respectively. Mesh concentration was applied near the inner autoclave walls to account for larger temperature and velocity gradients expected near solid walls, whereas a uniform mesh was used for the region surrounding the seeds [83]. Moreover, a fine grid was used in proximity of the baffle, where oscillatory flow was observed [83].

In their large eddy simulation of a laterally heated enclosure for crystal growth, Enayati et al. observe a closer agreement of an intermediately fine mesh with experimental results [31], which is unusual given that the accuracy of a numerical simulation generally improves for a finer mesh. They ascribe this effect to insufficient sampling resolution of the experimental data [31].

In the case of transient simulations, a discretization in time is also required. While the growth process itself is transient in nature (albeit slow), there is no evidence of whether fully stable temperature and flow fields develop under specific conditions. This is of practical interest for three reasons. Firstly, if a stable flow and temperature field exists during the main part of a growth run at constant set temperatures, numerical studies addressing temperature and flow field in this main part of the growth run can be investigated by steady computations. Secondly, the stability of the flow and temperature field is also expected to influence the growth process because it will affect the stability of mass transport and the driving force for crystallization. Thirdly, the ability to obtain a steady solution can have practical implications if certain software functionalities are available only for steady or only for transient studies in commercial CFD software. Several numerical studies have shown that the flow shows oscillatory features, at least under certain conditions [26,27,29,50]. More recently, an experimental study has provided experimental evidence that significant fluctuations of local fluid temperature can indeed occur, and the experimental data do not show a strictly oscillatory behavior of these temperature fluctuations [43]. However, these numerical and experimental occurrences of flow instability do not necessarily imply that stable conditions never exist. Masuda et al. have presented steady-state temperature and flow fields for an ammonothermal system with normal solubility [25]. Jiang et al. conducted transient calculations with a variety of open/space ratios of a ring baffle and found that there are oscillations in volume flux [27]. According to their results, the amplitude of oscillations varies with the open/space ratio, and thus this ratio can be used to establish comparatively stable flow conditions [27], though none of them appears to be fully stable. It should be noted that there are two qualitatively different ways in which the flow can be transient under quasi-steady conditions (we define quasi-steady conditions by stable wall temperatures at the control thermocouple locations over an extended period of time). The first possibility is that flow speed fluctuates but flow directions are stable. The second possibility is that flow speeds oscillate around zero, thus changing the direction of flow at a specific location over time. Such oscillations have been reported by Chen et al. [42]. From a general CFD viewpoint, C. Hirsch recommends the use of transient equations unless one can be assured that the flow will remain steady ([63], p. 142f), at least if there is no major advantage in using steady equations.

3.6. Results

In the following, selected results from the literature on flow and temperature field simulations of ammonothermal growth reactors are reviewed. Note that the intent is rather to draw the reader's attention to likely relevant aspects than to provide a complete overview.

Chen et al. have shown that the position of the nutrient strongly affects the temperature field inside the autoclave and point out that the inversed positioning of dissolution and growth zones for normal and retrograde solubility causes distinctly different temperature fields [42] (not shown here). Moreover, Chen et al. investigated the fluid flow in the vicinity of the baffle for the normal solubility case for different inner diameters of the autoclave [52]. The setup considered by Chen et al. is depicted in Figure 5a. The temperature field is shown in Figure 5b. The average flow velocities in the center hole of the baffle and in the ring-shaped gap between the baffle and autoclave wall are shown in Figure 5c (presumably averaged over the respective cross-section areas). As it can be seen from c, average flow velocities are subject to major fluctuations over time, which vary with the inner diameter of the autoclave [52] and the open-to-space ratio [42]. In spite of the fluctuations in flow velocity, the flow largely remains upward in the center hole and downward in the ring gap [52]. However, Chen et al. also report a case in which the flow direction in the center hole alternates over time [42].

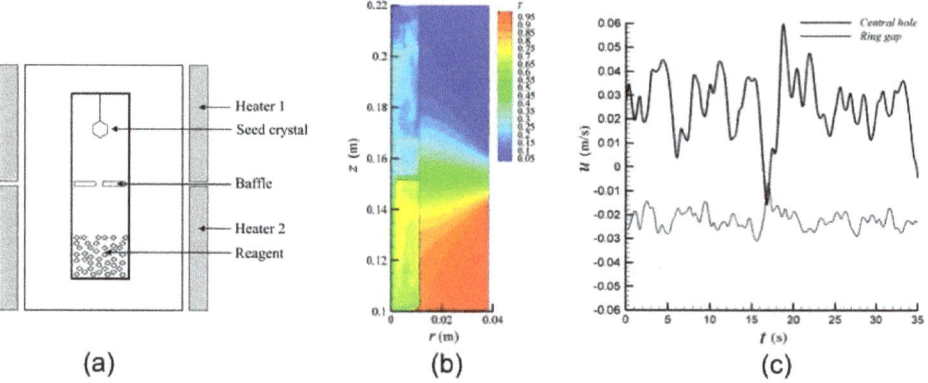

Figure 5. Setup (**a**) and simulation results (**b**,**c**) by Chen et al. for a reactor with an inner diameter of 2.22 cm and baffle opening of 10%: dimensionless temperature field (**b**) and average velocity in the central hole and ring gap (**c**). Reprinted with permission from Springer Nature Customer Service Centre GmbH: Springer Nature Research on Chemical Intermediates [52], Copyright 2011.

Gaps between autoclave wall and baffle as well as between autoclave wall and nutrient have been neglected in some numerical studies; however, as pointed out by Mirzaee et al. the presence of such a gap is a fact of experimental research [29]. Moreover, Masuda et al. cite their empirical knowledge that a small gap is necessary between autoclave wall and baffle [25]. Mirzaee et al. investigated the influence of the gap between autoclave wall and nutrient and concluded that it induces a circulating flow through the nutrient, which lowers the temperature difference but greatly enhances the transport of Ga-containing solutes out of the porous medium [29]. Although the solubility data, flow model and boundary conditions used by Mirzaee may not yet be accurate, this conclusion appears highly plausible. The importance of the gap between autoclave wall and baffle has previously been pointed out by Pendurti et al. who investigated transport in ammonothermal growth [30]. It can be concluded that although thin gaps are undesirable from a meshing viewpoint, it is essential to include the gaps between the autoclave wall and baffle as well as between the autoclave wall and nutrient basket.

Masuda et al. have investigated how the flow field and temperature distribution in the growth zone change with increasing crystal size [25]. They find a transition from one flow pattern to another flow pattern at a specific crystal radius and observe that this change in flow pattern also leads to changes in the temperature distribution in and around the crystal (see Figure 6). Similarly, Mirzaee et al. have also reported significant effects of the geometrical changes of growing crystals on the flow field [29].

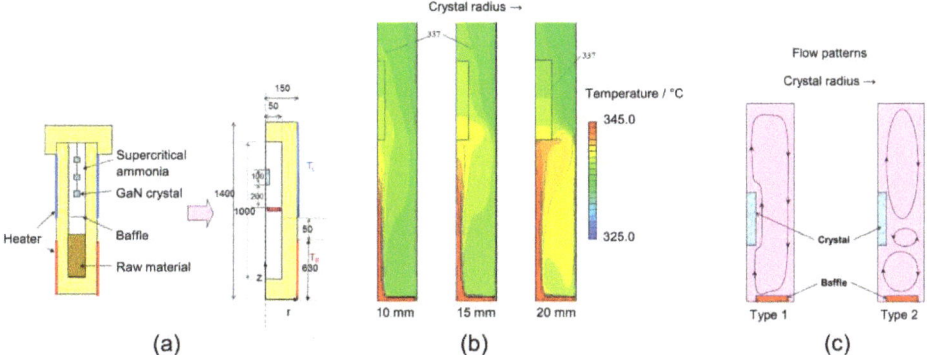

Figure 6. Effects of crystal size increase in a 2D simulation by Masuda et al. (**a**) Considered experimental setup and chosen axisymmetric model thereof, (**b**) temperature distribution in the vicinity of the seed and above the baffle, (**c**) schematic of the two types of convection patterns that were found to develop for crystal radii below and above about 15 mm, respectively. Reprinted with permission from [25]. Copyright (2016) The Japan Society of Applied Physics.

Erlekampf et al. as well as Enayati et al. have concluded that the fluid flow is three-dimensional even for an axisymmetric simulation domain [26,31]. However, it remains unknown how much those fluctuations on a small timescale affect the results of a growth run. Either way, one can expect three-dimensional flow to have practically relevant effects in the vicinity of the not axisymmetric seeds, where effects of the 3D geometry of the seeds are likely to also affect the time-averaged flow and temperature fields.

In practice, another relevant lack of axial symmetry may arise from the imprecise alignment of the axis of the thermal field of the furnace and the axis of the autoclave. Such alignment issues have been studied for hydrothermal growth by Li et al. who studied an industrial-scale autoclave [84]. Based on measured temperature deviations of around 2 K, Li et al. estimated the heat flux on the outer autoclave wall to vary by as much as 10% [84]. Li et al. conclude that the circumferential temperature variations in an industrial hydrothermal growth reactor are sufficient to establish an asymmetric flow [84]; however, they do not report whether the flow in a 3D model with symmetric boundary conditions and axial symmetry is symmetric or not.

Recently, a study on the effects of thermal boundary conditions was conducted by Schimmel et al. [57]. Following the boundary condition approach illustrated in Figure 4a, a global simulation of the temperature and flow fields was used to determine a realistic temperature distribution at the outer autoclave wall [57]. This temperature distribution was then used as a thermal boundary condition for a less complex model using the approach visualized in Figure 4d. For comparison, the otherwise same model was also solved with heater-long fixed temperatures and otherwise adiabatic walls (corresponding to the approach shown in Figure 4c. The resulting temperature and flow fields are depicted in Figure 7. Two aspects are of general importance. Firstly, the outer wall temperature distributions shown Figure 7a,c are remarkably different from the temperature distribution shown in Figure 7b. This indicates that at least for the studied ammonothermal setup with two heating zones and an uninsulated autoclave head, the wall temperature distribution differs significantly from the idealization of heater-long fixed temperatures and otherwise adiabatic walls [57]. Secondly, it should be noted that the modified thermal boundary conditions do have a pronounced effect on the temperature distribution and fluid flow field in the autoclave cavity [57]. For achieving simulations with reasonable agreement with experimental results, it therefore appears crucial to use more realistic thermal boundary conditions than those that have most commonly been applied in the literature [57]. Figure 5d–f show the corresponding flow fields in the cavity.

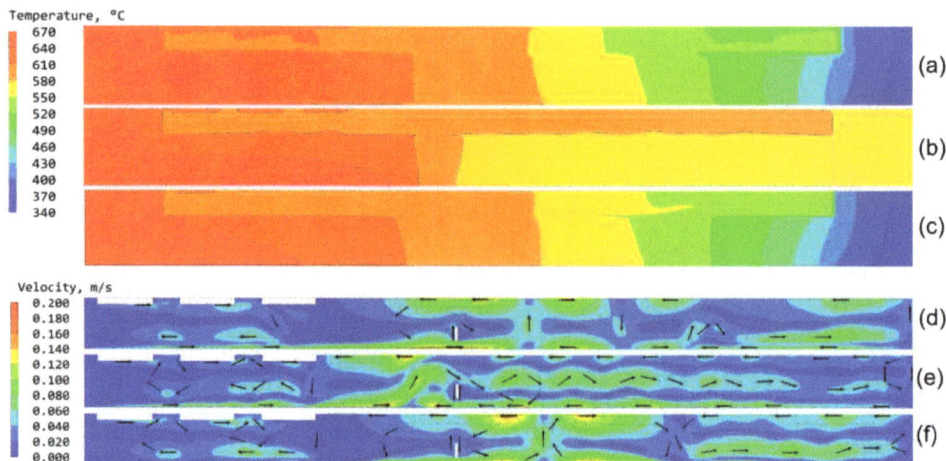

Figure 7. Comparison of the effects of different thermal boundary conditions applied to the outer autoclave walls. (**a**,**d**) show temperature and flow fields of a global field simulation including the furnace and its surroundings, (**b**,**e**) temperature and flow fields with heater-long fixed wall temperatures and otherwise adiabatic walls. (**c**,**f**) have been obtained using wall temperature distributions extracted from a global field simulation including the furnace and its surroundings with a simulation domain limited to the autoclave. In all cases, three seed crystals and a ring-shaped baffle were considered in an axisymmetric 2D simulation. Major arrows are normalized and serve as a guide to the eye for better visibility. Reprinted from [57].

4. Simulations of the GaN Crystal Growth Process

Pendurti et al. have presented steady-state growth rates for the case of transport-limited growth, on the basis of early solubility data for ammonobasic growth [30]. They note that their predicted transport-limited growth rates are in the order of hundreds of microns per hour, whereas experimentally observed growth rates are in the order of hundreds of microns per day, and ascribe this deviation to neglecting dissolution and growth kinetics [30]. The fact that the grown crystals usually show formation of facets [20,85,86] and the observation that nutrient loss mass flux does not covary with seed mass flux [44] both imply that seeded GaN growth is likely surface-reaction-limited in state-of-the-art growth processes. It is interesting that this holds for both ammonobasic and ammonoacidic process variants, suggesting that there might be a common limiting factor for the growth kinetics. Griffiths et al. pointed out that (at least for ammonobasic growth with Na mineralizer) polar growth rates dominate at low growth zone fluid temperature, whereas nonpolar growth rates begin to approach polar growth rates as temperature increases [44]. However, transport limited growth appears to occur in industrial hydrothermal growth of quartz crystals [84]. Regardless of the limiting factor for growth rates, it should be noted that the simulation by Pendurti et al. is a transient simulation over many hours, as they simulated how etch-back transitions into growth and how stable growth rates are approached [30]; however, it does not constitute a complete simulation of the growth process (such as with a time-dependent temperature program at the autoclave walls). Rather, a transient simulation is used to study how a steady temperature, fluid flow, and concentration field develop, starting from initial conditions, and which steady-state transport-limited growth rate is established as a result.

Mirzaee et al. however, have presented an actual process simulation over a growth time of 100 h and including the advancement of the crystal-solution interface [29], which represents a significant advancement compared to earlier studies that had purely focused on the quasi-steady phase of a growth run without accounting for geometrical changes of internal solids. Specifically, fluid flow, heat transfer, GaN metastable phase transport, dissolution and crystallization rates, and crystal formation have been included in the model by Mirzaee et al. [29]. In agreement with the experience of the first author of the present paper, they state that the needed time step size and thus computation times depend critically on the stability of the fluid flow inside the autoclave [29]. Per 24 h of the growth process, Mirzaee et al. needed CPU times from 0.5 to 3 days, depending on the stability of fluid flow, in spite of using a 2D model (on an Intel (R) 3.52 GHz CPU Seven desktop) [29]. Thus, computation times for the 100 h long process that they considered must have varied from 2 to 12.5 days.

5. Approaches to Validation

At present, there is a lack of experimental validation of simulations that study the ammonothermal growth of GaN, and studies that include only temperature and flow fields are no exception. This issue originates from the difficulty of experimental access to the interior of the autoclave, which represents a critical issue because it affects numerical simulations not only through the lack of validation data but also through a lack of accurate knowledge on physical and chemical quantities of the system. Over the last decade, significant progress has been made in the area of in situ monitoring technology for ammonothermal autoclaves. Alt et al. not only demonstrated local internal temperature measurements using thermocouples but also developed an optical cell suitable for optical measurements such as video optical observations and spectroscopical measurements [87]. Optical cells were further developed by Steigerwald et al. and used for investigating the decomposition of ammonia and ammonobasic solutions under supercritical conditions via UV/vis and Raman Spectroscopy [88]. Besides the optical applications, optical cells have also enabled in situ X-ray imaging with moderate X-ray energies, which has been applied primarily for studying dissolution kinetics and solubility of GaN under a variety of experimental conditions [56,89]. Due to the good X-ray transparency of the applied window materials [90,91], the in situ X-ray imaging method developed by Schimmel et al. allows even for tracking local concentration changes of Ga-containing solutes [39]. More recently, Schimmel et al. have also demonstrated that internal temperature measurements contain additional information beyond fluid temperatures, namely information on the stability of fluid flow and on chemical reactions associated with enthalpy changes [43]. Despite those encouraging developments, a few difficulties for their use for validating numerical simulations remain. Firstly, it would be highly desirable to validate temperature and flow fields; however, both fields are notoriously difficult to obtain with a good special resolution, for the entire interior of an autoclave and without influencing the flow by the introduction of measurement devices such as thermocouples. The latter issue can in principle be addressed by introducing the thermocouples also in the numerical simulation; however, it can require significant adaptation of the mesh, and it becomes problematic for rotational symmetry if thermocouples are to be placed away from the centerline of the autoclave.

Enayati et al. [31] and Moldovan et al. [28] presented an interesting approach of validation via the design of a geometrically and dynamically similar experimental setup that allows for particle image velocimetry-based flow visualizations and thus yields actual flow fields. Moldovan et al. found the flow to be oscillatory in nature but with steady time-averaged patterns [28]. They used the standard deviation of fluid temperatures to assess the magnitude of fluctuations at different locations, and found the fluctuations to be most pronounced in the mixing region between the hot and cold sections [28]. However, it should be noted that two uncertainties remain in this case. Firstly, it is difficult to ensure similarity of ammonothermal growth setup and experimental model due to the limited knowledge of the physics of the ammonothermal system, in particular its fluid properties under process conditions. Secondly, while the model is designed to provide similar thermal boundary conditions as the numerical model, there appears to be a high probability that those boundary conditions may differ significantly from those of the actual ammonothermal growth experiment (see Section 3.6. and [57]). In the author's opinion, the most important advantage of using a physical model reactor with particle image velocimetry is that it allows for visualizing large parts of the flow field. Although visual access could in principle be implemented for growth reactors, any method that relies on nonmetallic windows (see e.g., [87,88]) will remain restricted in the area of view for the typical pressure and temperature conditions of ammonothermal processes, which represents a severe limitation specifically for flow visualization. The confidence in the similarity of a physical model system would benefit from more detailed knowledge on the outer wall temperatures of the ammonothermal autoclave and on the fluid properties of the ammonothermal fluid mixture. The approach of using time-averaged patterns and local standard deviations of fluid temperatures should also be transferable to in situ validation experiments using internal thermocouples under actual process conditions, such as the one recently presented by Schimmel et al. [43].

Mirzaee et al. state that their code underwent validation by a variety of means as follows. Firstly, the code was tested on natural convection problems for which experimental validation data exist [29], albeit using fluids with better-known properties such as glycerin and water [92]. While being perfectly reasonable for validating most aspects of the natural convection problem, this approach does not represent a complete validation because the uncertainty about fluid properties of the solute-containing supercritical fluid mixture under ammonothermal process conditions is not eliminated. In addition, one also needs to keep in mind that the calculation of supersaturation is based on early experimental solubility data published by Wang et al. in 2006 [93] and that the solubility of GaN under various process conditions remains an active area of research and scientific discussion [56,94]. A second step of validation taken by Mirzaee et al. is using it on a flow pattern in a retrograde solubility ammonothermal crystal growth system that had been presented by Chen et al. [42]. The data published by Chen et al. [42], however, represent a numerical study of unknown accuracy in itself and therefore do not allow for a complete validation either. For validating the interface advancement, Mirzaee et al. utilize the experimentally validated problem of the deformation of the solid–fluid–gas interface in water upon the entry of a solid sphere into water [95]. While the authors of this review wish to encourage all reasonable validation approaches even if yielding only a partial validation, neither one of the validation approaches conducted by Mirzaee et al. can fully confirm the accuracy of their simulation, which is largely due to a lack of knowledge on fundamental aspects such as fluid properties and solubility.

To the knowledge of the authors, no simulation of the entire ammonothermal GaN growth process with comprehensive experimental validation exists to date. Such simulations appear to be limited to much more well-investigated crystal growth processes such as Czochralski growth of silicon, and even for this extremely well-established growth process, they are still an active area of research [96]. It is furthermore interesting to note that even for such well-established crystal growth processes, validation of simulations using physical models also remains an active area of research, as indicated by a recent review by Dadzis et al. [97]. This indicates that even with comparatively well-established material properties, it is not always clear which simplifying assumptions yield the right balance of computational speed and accuracy. The proper choice of assumptions, however, is crucial for developing a simulation that is of full practical use for engineering a crystal growth process.

6. Open Questions That May Affect the Accuracy of Simulation Results

In case of ammonothermal crystal growth and the conditions in respective high-pressure reactors, there are a variety of possible causes for significant deviations of simulation results from experimental observations. Some potentially major sources of uncertainty are discussed in this section. Unfortunately, validation by comparing simulation results to experimental data is feasible only to a limited extent because of the technical difficulty of measuring temperatures, flow velocities, and other simulated quantities during ammonothermal experiments. In addition, while hydrothermal crystal growth benefits from synergies with geological research, this is not the case for the ammonothermal growth of nitrides. Hence, the thermodynamics, kinetics, and fluid properties in ammonothermal systems are still underexplored. Thus, there is a particularly high risk of making inadequate assumptions without noticing it.

In the following, potentially critical issues will be discussed. Related experimental observations will also be mentioned where available.

6.1. Fluid Properties

Ammonothermal crystal growth utilizes supercritical ammonia as a solvent. Pure substance properties will be illuminated first, followed by a discussion of possible deviations due to the presence of decomposition products of ammonia and solutes. Thermophysical data for pure ammonia are available up to 426.9 °C in a database of the National Institute of Standards and Technology (NIST) [98]. To give an overview of how different fill levels alter the pure substance properties of ammonia, Figure 8 shows density, pressure, specific heat capacity, dynamic viscosity, and thermal conductivity at the upper end of the temperature range for which data are available in the database by NIST [98]. The fill level refers to the volume fraction of the reactor filled with liquid ammonia at the boiling point of ammonia (−33.36 °C). It should be noted that even for the temperature range available in the database by NIST, basic properties are still updated occasionally as more accurate data become available [99]. Specifically, the thermal conductivity has recently been found to be about 6% lower than the respective database values [99].

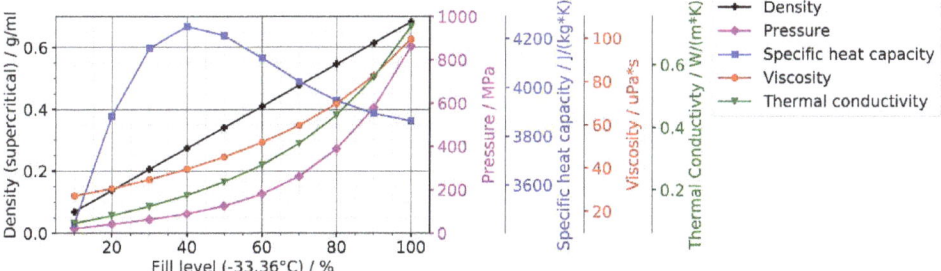

Figure 8. Properties of supercritical ammonia at 426.08 °C as a function of fill level (data from National Institute of Standards and Technology (NIST) database [98]).

The fill level represents a common quantity used by experimenters for obtaining a specific density of supercritical ammonia or a specific pressure at a specified mean temperature. Figure 9 shows the system pressure resulting from different fill levels as a function of temperature. Fill levels for ammonothermal growth of GaN are typically in the range of 30 to 70%, with those for ammonoacidic growth being lower than those for ammonobasic growth.

Since ammonothermal experiments employ temperatures typically up to 600 °C and sometimes even higher temperatures, extrapolated data are typically used [26,51]. An alternative approach is to simulate the fluid properties themselves. For instance, Masuda et al. have calculated liquid thermal conductivity and liquid viscosity of supercritical ammonia using the chemical process simulator VMGSim produced by Virtual Materials Group [25].

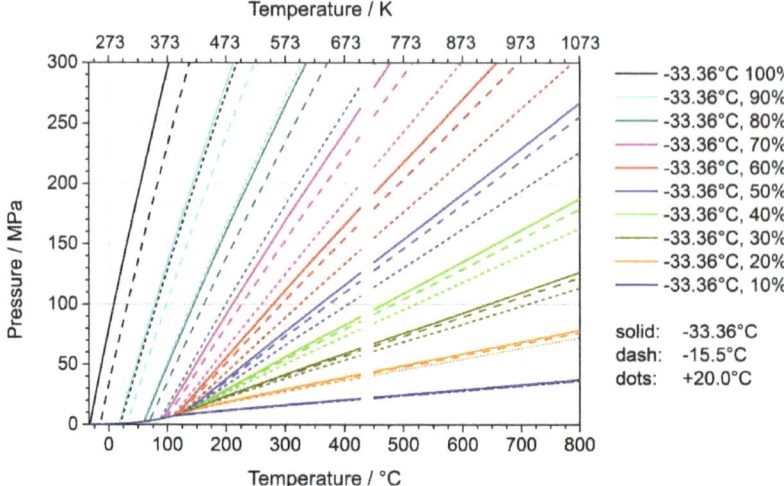

Figure 9. System pressure for ammonia in a closed system as a function of temperature for fill levels from 10 to 100%, with fill levels referring to three different reference temperatures (reprinted from [46]). The reference temperatures correspond to the boiling point, a typical temperature for the introduction of liquified ammonia using a pressurized system, and room temperature, respectively [46].

It should be noted that ammonothermal crystallization of GaN is conducted at rather unusual process parameters. Most applications of supercritical fluids, in particular extraction processes, utilize a parameter range that is characterized by a high isothermal compressibility [100]. Specifically, temperatures are typically in the range of 1.01 $T_c < T <$ 1.2 T_c, and pressures are usually in the range of 1.01 $p_c < p < 1.5\ p_c$ [100,101], with T_c and p_c representing the critical temperature and critical pressure of the fluid, respectively. For ammonia, this range would be 133.7 °C to 158.9 °C and 11.4 MPa to 17.0 MPa, which is far from the parameter range of ammonothermal growth of GaN [46]. This is illustrated in Figure 10b, which shows the pressure as a function of specific volume for different temperatures.

Given the existence of a much further developed but similar method, the hydrothermal growth of oxides, it appears reasonable to consider whether lacking data can be derived from the knowledge on that method. A comparison of exemplary process parameters for hydrothermal and ammonothermal crystal growth processes shows that the hydrothermal methods operate much closer to the critical temperature. However, both hydrothermal and ammonothermal crystal growth operate at pressures outside the typical application range for supercritical fluids, although this is more pronounced in the case of the ammonothermal method. The described comparison is substantiated by the data given in Table 1.

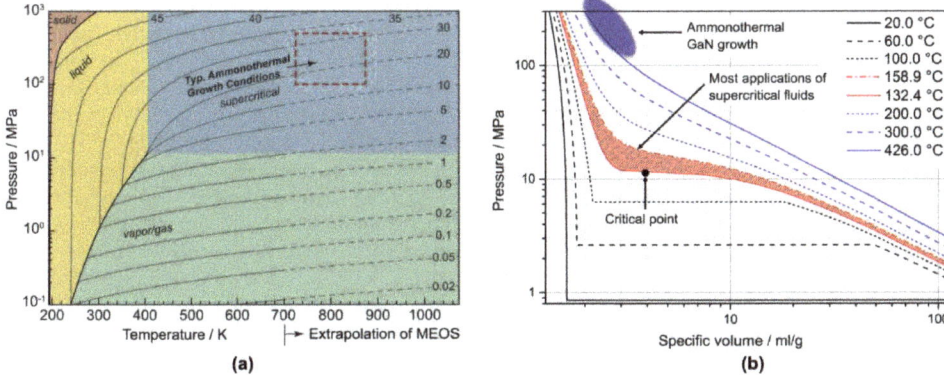

Figure 10. (**a**) Phase diagram of pure ammonia with contour lines of density (mol/l) and superimposed phases present at equilibrium. Calculated and extrapolated beyond 700 K using the reference multiparameter equation of state (MEOS) as provided by the National Institute of Standards and Technology (NIST). Reprinted from [102], Copyright 2016, with permission from Elsevier. (**b**) Pressure as a function of specific volume of ammonia for selected temperatures (reprinted from [46], data from NIST database [98]). The hatched area in red marks the typical application range of supercritical fluids whereas the dark blue region approximately indicates the parameter range of ammonothermal GaN growth.

Table 1. Process parameters of hydrothermal and ammonothermal growth. In the case of the ammonothermal method, the reports with the highest growth rates as of 2018 were chosen. The temperature dependency of solubility refers to the temperature range used in the growth process. T_{CZ} refers to the crystallization zone temperature, ΔT to the temperature difference between growth and dissolution zones. Reprinted from [46].

	Hydrothermal				Ammonothermal	
Material		Quartz		ZnO	GaN	
Process Route	Mineralizer-Free [103]	Low-Pressure Process [104]	High-Pressure Process [104]	[105]	Acidic [106]	Basic [3]
T_{CZ}/°C	445–500	345	360	300–430	625	575
T_{CZ}/T_c	1.19–1.34	0.92	0.96	0.80–1.15	4.73	4.35
ΔT/°C	25	10	25	10–20	50	30–45
Solubility	retrograde	normal		normal	retrograde	
p/MPa	60–110 [1]	70–100	100–150	70–255	80–150	250
p/p_c	2.71–4.98	3.17–4.52	4.52–6.79	3.17–11.54	7.08–13.27	22.12
Mineralizer	none	Na$_2$CO$_3$	NaOH	NaOH	NH$_4$F	Na
[0001] growth rate/µm/day	0.3–2	400	1000	300	410	344

[1] Based on the reported fill level.

In conclusion, it is not clear how accurate extrapolations and simulations of fluid data are, and thus, experimental data on fluid properties would significantly contribute to an increased confidence in the results of numerical simulations of ammonothermal reactors.

While extrapolation of the pure substance properties to higher temperatures may be reasonably accurate for pure ammonia, the question arises whether properties of pure ammonia are a reasonably accurate assumption. During ammonothermal growth, part of the ammonia decomposes, leading to a mixture of NH$_3$, N$_2$, and H$_2$ [102]. Pimputkar et al. have studied this decomposition reaction in a combined numerical and experimental approach to determine an accurate description for the equilibrium constant for the ammonia decomposition reaction as a function of pressure and temperature and verified it against experimental data [102]. The determined equilibrium constant as a function of inverse temperature is shown in Figure 11a. For selected fill densities, the calculated mole fractions of ammonia in equilibrium as a function of temperature are depicted in Figure 11b alongside experimental data. Depending on the materials of the pressure-bearing materials, hydrogen may leave the otherwise closed system by diffusion [102]. When considering the composition of the ammonothermal reaction medium during actual GaN growth experiments, it is important to be aware that different ammonothermal growth environments may cause vastly different kinetics of ammonia decomposition. In other words, any ammonia mole fraction from 1.0 down to the equilibrium value may be present at some point in time during a growth experiment, and there is also a possibility that equilibrium is never reached. The kinetics of ammonia decomposition can be expected to depend heavily on the presence or absence of materials that can act as a catalyst for the decomposition reaction. Specifically, Ni is known to catalyze ammonia decomposition [107]. GaN growth is typically conducted in autoclaves made from nickel-base [9,20,44,48] and sometimes molybdenum-base alloys [47]; however, the autoclave wall is not necessarily in direct contact with ammonia. In order to prevent corrosion of the autoclave wall, as well as in order to minimize the introduction of transition metal impurities, hermetically sealed [108] or pressure-balanced [3,109] liners or capsules of different, more corrosion-resistant materials are often applied. Depending on the mineralizer, platinum [9,40,106], silver [19], or molybdenum [70] are used as liner materials for bulk GaN growth. Besides the catalytic properties of the inner wall, there is also a possibility that the mineralizer itself may affect ammonia decomposition [46]. Given that chlorine is known to poison catalysts of ammonia synthesis and decomposition [110–112], NH_4Cl (and possibly further acidic mineralizers) might hinder ammonia decomposition [46]. In conclusion, it is not fully clarified how quickly ammonia decomposes under specific growth conditions of GaN. Consequently, it is not always clear which mole fractions of ammonia, nitrogen, and hydrogen should be assumed. To the author's knowledge, there are also no numerical studies that account for the presence of nitrogen and hydrogen. Moreover, the sensitivity of simulation results to changes in ammonia, nitrogen, and hydrogen mole fractions has not been investigated yet.

Moreover, solutes such as mineralizers and intermediates are essential components of ammonothermal reaction media. In situ measurements of Ga-intermediate concentrations and diffusive transport velocity indicate that there may be a significant increase in fluid viscosity when mineralizer and Ga-containing intermediates are present in the solution [39]. In situ measurements using a rolling ball viscosimeter have been shown to be feasible [113], but no viscosity data for typical solutions are available in the literature yet.

Besides the unknown influence of solutes on fluid viscosity, there is also uncertainty about the optical properties of the solution and hence about the relevance of heat transfer to and through the fluid via radiation. In simulations, the fluid is commonly assumed to be clear [29] and radiation is usually neglected. However, at least with NH_4Cl mineralizer, optical in situ measurements have shown that the optical transparency of the fluid decreases rapidly as temperature is increased, which is ascribed to increasing concentrations of solutes [87]. Though no measurements of transparency of infrared radiation have been reported, significant absorption of the fluid cannot be excluded and may be specific to the chemical species present.

An additional possibility is an altered heat capacity in the presence of solutes, as suspected by Alt et al. who performed temperature measurements with thermocouples directly in the fluid and compared experiments with and without NH_4Cl mineralizer [87]. However, Alt et al. do not comment on the reproducibility of their experiments, and the temperature deviations cannot unambiguously be assigned to a possible change of heat capacity.

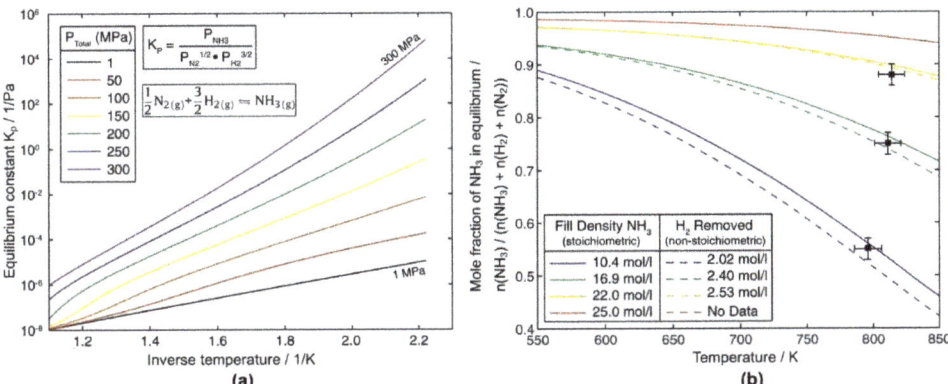

Figure 11. Data on ammonia decomposition at conditions relevant to ammonothermal GaN growth calculated by Pimputkar et al.: (**a**) equilibrium constant K_p for the ammonia decomposition reaction as a function of inverse temperature calculated for various total system pressures ranging from 1 to 300 MPa., (**b**) calculated equilibrium mole fraction of ammonia as a function of temperature and initial fill density, overlaid with three experimentally determined data points. Both reprinted from [46], Copyright 2016, with permission from Elsevier.

Last but not least, there is also a possibility that the ammonothermal reaction medium contains nanoparticles, as suspected in a study that combined experimental observations by in situ X-ray imaging with molecular dynamics simulations [39]. In recent years, nanofluids have been researched intensively as heat transfer liquids, as they often exhibit a strongly increased thermal conductivity with respect to the base fluid [114]. Moreover, thermophoresis plays a significant role in heat transfer and convection in nanofluids, as shown by J. Buongiorno [115] and studied further from a thermodynamics viewpoint by E. Bänsch [116]. Thermophoresis leads to a reduced concentration of particles in the proximity of hot walls, if they exhibit a significant thermal gradient to the fluid [115,116] (which can be expected in the case of autoclave walls, especially in the growth zone in ammonothermal growth). The locally reduced particle concentration in turn leads to a locally reduced viscosity; thus, thermophoresis enhances transport in the boundary layer [115,116]. Potential effects of nanoparticles in the solution have not been considered in any macroscale simulations so far.

6.2. Possible Relevance of Solutal Buoyancy

Since there are strongly deviating data on solubility of Ga (see Section 6.3, Solubility of the Metal), a possible contribution of solutal buoyancy can only be gauged with limited accuracy. Solutal buoyancy may have different effects for growth in the normal and retrograde solubility regimes, as the targeted transport direction of the metal is inversed with respect to gravity. A solutal contribution to convection should be more pronounced at low ammonia densities for a given Ga concentration; however, this may be compensated by the increase of solubility with increasing solvent density. For a rough estimate of whether solutal buoyancy can be expected to play a significant role, we attempt to quantify the range of local density differences under ammonothermal growth conditions of GaN based on thermal gradients and resulting differences in solute concentration. For local density differences of ammonia, we build on internal temperature measurements in an actual growth setup presented by Griffiths et al. [44]. According to this ammonobasic GaN growth study, growth becomes transport-limited at fluid density differences below 1.2 mol/L between dissolution and growth zone (based on extrapolation of NIST fluid density data of pure NH_3). For lack of data on Ga concentration differences under actual growth conditions, we utilize data on the solubility limit of Ga, which were obtained under ammonoacidic conditions using NH_4F mineralizer [39]. While the effect of solutal buoyancy may be smaller in reality (as there is likely no region in the autoclave with zero concentration of dissolved Ga), this consideration should be sufficient to elucidate whether solutal convection is likely to be of relevance. The results of this estimation are presented in Table 2. Accordingly, solutal buoyancy may very well have a significant impact, as the density increase due to Ga-containing solutes reaches the same order of magnitude as the density difference of pure ammonia that is induced by thermal gradients.

Table 2. Estimate of temperature-induced density differences (labeled "supercritical NH_3") between growth and dissolution zone, and solute-induced density differences (labeled "dissolved Ga").

	Supercritical NH_3	Dissolved Ga
Density difference/mol/L	1.2–5.8 [44]	0.5 [39]
Density difference/g/L	20.4–104.6	34.9

6.3. Solubility of the Metal

For simulations that include mass transport of the metal (which has already been attempted [29]), an additional uncertainty arises from the limited and deviating data on the solubility of the metal. An overview of available data and investigated parameter ranges is given in Table 3. Although there are several reports on GaN solubility under ammonothermal conditions, the different reports cover widely scattered but not nearly comprehensive parameter ranges regarding important variables such as solvent density or pressure, mineralizer concentration, mineralizer substance, and temperature. Different methods have been applied, aiming at obtaining reliable quantitative solubility data; however, data do not appear to be in good agreement yet, and there is an ongoing discussion on causes of measurement errors [56,89,94,117]. There are two aspects that are especially important to numerical simulations of the growth process: the temperature dependency of solubility and its absolute magnitude. The existing solubility studies have mostly focused on limited parameter ranges to function as guidance for experimental research in individual laboratories, for which the semi-quantitative change of solubility with certain parameters such as temperature or mineralizer concentration is most important.

The temperature dependency of solubility governs the (super-)saturation field and is therefore critical for the driving forces for both dissolution and crystallization. If only a single solute species dominates solubility for a range of temperature, the heat of solution is directly related to the temperature dependency of solubility via van 't Hoff equation, and for such cases, the heat of solution has been determined experimentally [9,117,118].

The absolute magnitude of solubility for a given set of parameters is also relevant because it governs the concentration field. The concentration field together with the flow field determines the mass transport by diffusion and convection. In addition, the absolute magnitude also affects the question of whether solutal buoyancy plays a relevant role or not.

It remains to be said that a comprehensive database of solubility would be valuable, especially for numerical studies of the growth process. In such a database, solubility should be available as a function of solvent density, mineralizer concentration (or possibly acidity or alkalinity of the solution), mineralizer substance, temperature, and possibly pressure. In hydrothermal systems (studied in geological contexts), the solubility of different minerals is known to depend either primarily on acidity/alkalinity or primarily on pressure and temperature, and solubilities have also been studied through numerical modeling [119]. A combination of further clarification of solution chemistry, numerical modeling, and experimental validation of selected conditions may eventually provide comprehensive data on ammonothermal solutions with sufficient accuracy for the use in growth process modeling. Both experimental and numerical investigations in the context of ammonothermal solution chemistry have recently been conducted [38,120,121] and represent important first steps in this direction.

Table 3. Overview of data range of GaN solubility under ammonothermal conditions (reprinted from [46] with added data from new publication [117]). Values marked with an asterisk have been converted to the respective unit. For normalization to volume, the volume of the reactor was used.

Mineralizer	Experimental Conditions	Range of Solubility Data	Reference
NH_4Cl/NH_4I mixture	450–550 °C (external) 96–102 MPa 0.42–0.51 mmol NH_4X/mL (*) 100 h	0.048–0.052 mol GaN/mol NH_4X (*) 0.15–1.2 mol% 0.42–0.47 mmol/mL (*)	D. Tomida [117]
NH_4Cl/NH_4Br mixture	450–550 °C (external) 96–103 MPa 0.40–0.51 mmol NH_4X/mL (*) 100 h	0.11–0.12 mol GaN/mol NH_4X (*) 0.35–1.23 mol% 0.40–0.51 mmol/mL (*)	D. Tomida 2018 [117]
NH_4Cl	200–550 °C (external) 67.7–100.9 MPa 0.33–3.30 mmol NH_4Cl/mL (*) 120 h	0–2.4 mol GaN/mol NH_4Cl 0–7.04 mol% (*) 0–7.92 mmol/mL (*)	D. Ehrentraut 2008 [122]
	420–600 °C (external) 55–150 MPa 0–4.04 mmol NH_4Cl/mL (*) 100 h	up to 0.41 mol GaN/mol NH_4Cl (*) 0.04–5.47 mol% 0.01–1.65 mmol/mL (*)	D. Tomida 2010 [118]
NH_4F	486–572 °C (internal) 16–175 MPa 0.76 mmol NH_4F/mL Until observation of saturation	0–0.11 mol GaN/mol NH_4F 0–1.03 mol% 0–0.08 mmol/mL	S. Schimmel 2017/2018 [56]/[46]
Na	415–650 °C (internal) 200 MPa 14.13–21.89 mmol Na/mL (*) 45–316 h	0.00017–0.00122 mol GaN/mol Na (*) 0.02–0.12 mol% 0.07–3.45 mmol/mL (*)	S. Griffiths 2016 [94]
$NaNH_2$	450–650 °C (external) 76 ± 12 MPa 0.14 mmol $NaNH_2/mL$ (*) 120 h	up to 0.16 mol GaN/mol $NaNH_2$ (*) up to 2.44 mol% (*) up to 0.02 mmol/mL (*)	T. Hashimoto 2007/2011 [123,124]
NaN_3	396–538 °C (internal) 259–268 MPa 0.65 ± 0.07 mmol NaN_3/mL Until observation of saturation	0.02–0.05 mol GaN/mol NaN_3 0.04–0.15 mol% 0.01–0.04 mmol/mL	S. Schimmel 2017/2018 [56]/[46]

6.4. Dissolution and Growth Kinetics

Regarding growth kinetics, Griffiths et al. have investigated growth kinetics in a kinetically limited ammonobasic regime and determined activation energies by Arrhenius analysis of the temperature dependency of seed mass flux [44]. Dissolution kinetics appear to be more difficult to analyze based on growth runs. Griffiths et al. mention the complication that source loss flux includes not only contributions to seeded growth but also contributions to parasitic deposition and Ga loss e.g., to autoclave walls [44]. In addition, in our view, it appears difficult to distinguish dissolution limited by surface kinetics from dissolution limited by transport in a growth experiment. In a methodically different approach in the context of solubility studies, dissolution kinetics of GaN under selected ammonothermal conditions have been investigated by in situ X-ray imaging [46,56,89]. While this yields direct, nearly real-time observation, the sluggish thermal response of the thermal mass of the autoclave and the difficulty of distinguishing kinetically and transport-limited regimes so far have prevented the extraction of kinetic parameters such as activation energies from in situ X-ray monitoring results. Altogether, data on dissolution and growth kinetics in ammonothermal GaN growth remain sketchy and do not yet provide a sufficient basis for growth process modeling for most experimental parameters.

7. Conclusions and Outlook

In conclusion, progress in the following three areas is required before a fully trustworthy numerical simulation with known accuracy and known limitations can be realized. Firstly, a variety of physical and chemical questions need to be addressed. Secondly, technical issues of numerical simulations need to be resolved. Thirdly, validation data need to be obtained via in situ measurement techniques and physical model systems. In particular, sound knowledge on fluid properties of the respective solute-containing mixtures under process conditions (or, alternatively, a proof of negligible deviation in properties from pure substance data in the available temperature range) would significantly increase the confidence in the results of both numerical simulations and physical models.

While establishing numerical simulations of known accuracy represents a great challenge, one should keep in mind that once those obstacles have been overcome, numerical simulations could become an incredibly useful tool for the further development of ammonothermal crystallization processes. Besides facilitating scale-up in the industrial use of the method, numerical simulations also hold the potential to achieve comparability of results from laboratories around the world through the sharing of internal experimental conditions. This could tremendously speed up development, especially in collaborative research. At least within collaborations, data exchange is often feasible, but there is often a lack of comparability or knowledge of the internal process conditions. Since experimental setups often cannot follow identical designs due to different requirements of each collaborative project or measurement technique, reliable numerical simulations would be extremely helpful for comparing or combining results from different experimental setups in a meaningful way.

Given the number of unknowns, validation of simulations is deemed to be essential for establishing reliable simulations of known accuracy. A feasible approach for obtaining rather comprehensive validation data for numerical simulations, including mass transfer of GaN is seen in the application of high-energy computed tomography [125] in conjunction with internal temperature measurements [43,44] while using outside wall temperature measurements for heater control. It is, however, important to note that a transient simulation, ideally even in 3D, would be necessary in order to take full advantage of such an approach for validation. The reason is that the information that becomes accessible by computed tomography for the entire autoclave volume is the distribution of solid GaN as a function of time. Therefore, the obtainable primary information is essentially validation data for a transient simulation of the entire growth process. Due to the number of unknowns as well as the need for balancing computational cost and accuracy, it is entirely possible that the agreement of simulation results and experimental validation data will initially be mediocre or possibly even poor. It would then be necessary to identify those unknowns that are likely to cause the observed deviations and to address the underlying issues (for instance by determining fluid properties). Hereby, the accuracy of simulations could incrementally be improved until a sufficient agreement of experimental validation data and numerical model is achieved.

While a physics-based simulation of the entire growth process will clearly be complex, different opportunities for speedup exist or are likely to become available in the foreseeable future. While computation times will likely remain an issue on workstation PCs, one should keep in mind that access to advanced computing resources such as supercomputers is still on the rise. In addition, recent progress in the area of artificial intelligence is increasingly being utilized for novel approaches in the field of computational fluid dynamics. This includes the use of deep convolutional neural networks for determining closure terms for spanwise-averaged Navier–Stokes equations, which allows us to account for 3D turbulence effects at a greatly reduced computational cost [126]. Specifically in the field of crystal growth, machine learning has already successfully been applied for speeding up numerical modeling of supersaturation and flow field in SiC solution growth using the results of physics-based simulations as training data for machine learning [127]. Integrated approaches utilizing a combination of machine learning and knowledge of physical laws (for example, physics-informed neural networks) are also increasingly researched for a variety of applications and can make machine learning more data-efficient [128]. In the long run, there is also a prospect of a disruptive expansion in available computing power through the advent of quantum computing, which may eventually provide tremendous speedup also for computational fluid dynamics and multiphysics simulations [129,130].

Once sound knowledge of the internal growth conditions has been obtained, it may also become feasible to utilize numerical simulations to improve the understanding of defect formation and impurity incorporation. For instance, the effects of inhomogeneous impurity incorporation, pressure and temperature changes, and other causes of stress may be worth investigating. Such studies have already yielded instructive insights for mitigating defect formation in more mature crystal growth techniques, such as Hydride Vapor Phase Epitaxy (HVPE) of GaN [131] and Physical Vapor Transport (PVT) growth of SiC [132].

Author Contributions: Conceptualization, S.S. and D.T.; methodology, S.S.; formal analysis, S.S.; investigation, S.S; resources, S.S., D.T., Y.H., and H.A.; data curation, S.S.; writing—original draft preparation, S.S.; writing—review and editing, S.S., D.T., T.I., Y.H., S.C., and H.A.; visualization, S.S.; supervision, D.T., Y.H., H.A., and S.C.; project administration, S.S., D.T., Y.H., and H.A.; funding acquisition, S.S., D.T., Y.H., and H.A. All authors have read and agreed to the published version of the manuscript.

Funding: This research was funded by JAPAN SOCIETY FOR THE PROMOTION OF SCIENCE (JSPS), grant number P19752 (Postdoctoral Fellowships for Research in Japan (Standard)). This research is supported by the MEXT "Program for research and development of next-generation semiconductor to realize energy-saving society" Program Grant Number JPJ005357.

Institutional Review Board Statement: Not applicable.

Informed Consent Statement: Not applicable.

Data Availability Statement: Data sharing not applicable.

Acknowledgments: The authors would like to thank Mitsubishi Chemical Corporation and The Japan Steel Works, Ltd. for their continuous support. The first author (S.S.) would also like to thank collaborators from previous ammonothermal projects, including, but not limited to, Peter Wellmann, Rainer Niewa, Wolfgang Schnick, Eberhard Schlücker, Elke Meissner, Thomas G. Steigerwald, Dietmar Jockel, and last but not least Siddha Pimputkar for fruitful discussions on various aspects of ammonothermal crystal growth. Such discussions have contributed to gaining a more comprehensive understanding of existing limitations of knowledge that may affect the accuracy of numerical simulations. In addition, the first author (S.S.) would like to thank the Alexander von Humboldt-Foundation for the nomination for the JSPS fellowship.

Conflicts of Interest: The authors declare no conflict of interest.

References

1. Juza, R.; Jacobs, H.; Gerke, H. Ammonothermalsynthese von Metallamiden und Metallnitriden. *Berichte der Bunsengesellschaft für Phys. Chemie* **1966**, *70*, 1103–1105. [CrossRef]
2. Dwiliński, R.; Wysmołek, A.; Baranowski, J.; Kamińska, M.; Doradziński, R.; Garczyński, J.; Sierzputowski, L.; Jacobs, H. GaN Synthesis by Ammonothermal Method. *Acta Phys. Pol. A* **1995**, *88*, 833–836. [CrossRef]
3. Pimputkar, S.; Kawabata, S.; Speck, J.S.S.; Nakamura, S. Improved Growth Rates and Purity of Basic Ammonothermal GaN. *J. Cryst. Growth* **2014**, *403*, 7–17. [CrossRef]

4. Bao, Q.; Saito, M.; Hazu, K.; Furusawa, K.; Kagamitani, Y.; Kayano, R.; Tomida, D.; Qiao, K.; Ishiguro, T.; Yokoyama, C.; et al. Ammonothermal Crystal Growth of GaN Using an NH 4 F Mineralizer. *Cryst. Growth Des.* **2013**, *13*, 4158–4161. [CrossRef]
5. Zajac, M.; Kucharski, R.; Grabianska, K.; Gwardys-bak, A.; Puchalski, A.; Wasik, D.; Litwin-Staszewska, E.; Piotrzkowski, R.; Z Domagala, J.; Bockowski, M. Basic Ammonothermal Growth of Gallium Nitride–State of the Art, Challenges, Perspectives. *Prog. Cryst. Growth Charact. Mater.* **2018**, *64*, 63–74. [CrossRef]
6. Key, D.; Letts, E.; Tsou, C.-W.; Ji, M.-H.; Bakhtiary-Noodeh, M.; Detchprohm, T.; Shen, S.-C.; Dupuis, R.; Hashimoto, T. Structural and Electrical Characterization of 2" Ammonothermal Free-Standing GaN Wafers. Progress toward Pilot Production. *Materials* **2019**, *12*, 1925. [CrossRef]
7. Li, T.; Ren, G.; Su, X.; Yao, J.; Yan, Z.; Gao, X.; Xu, K. Growth Behavior of Ammonothermal GaN Crystals Grown on Non-polar and Semi-polar HVPE GaN Seeds. *CrystEngComm* **2019**, *21*, 4874–4879. [CrossRef]
8. Ehrentraut, D.; Pakalapati, R.T.; Kamber, D.S.; Jiang, W.; Pocius, D.W.; Downey, B.C.; McLaurin, M.; D'Evelyn, M.P. High Quality, Low Cost Ammonothermal Bulk GaN Substrates. *Jpn. J. Appl. Phys.* **2013**, *52*, 08JA01. [CrossRef]
9. Wang, B.; Callahan, M.J. Ammonothermal Synthesis of III-nitride Crystals. *Cryst. Growth Des.* **2006**, *6*, 1227–1246. [CrossRef]
10. Dwiliński, R.; Doradziński, R.; Garczyński, J.; Sierzputowski, L.P.; Puchalski, A.; Kanbara, Y.; Yagi, K.; Minakuchi, H.; Hayashi, H. Excellent Crystallinity of Truly Bulk Ammonothermal GaN. *J. Cryst. Growth* **2008**, *310*, 3911–3916. [CrossRef]
11. Suihkonen, S.; Pimputkar, S.; Sintonen, S.; Tuomisto, F. Defects in Single Crystalline Ammonothermal Gallium Nitride. *Adv. Electron. Mater.* **2017**, *3*. [CrossRef]
12. Amano, H. Growth of GaN Layers on Sapphire by Low-Temperature-Deposited Buffer Layers and Realization of p-type GaN by Magesium Doping and Electron Beam Irradiation (Nobel Lecture). *Angew. Chemie Int. Ed.* **2015**, *54*, 7764–7769. [CrossRef] [PubMed]
13. Kucharski, R.; Sochacki, T.; Lucznik, B.; Bockowski, M. Growth of Bulk GaN Crystals. *J. Appl. Phys.* **2020**, *128*. [CrossRef]
14. Paskova, T.; Evans, K.R. GaN Substrates-progress, Status, and Prospects. *IEEE J. Sel. Top. Quantum Electron.* **2009**, *15*, 1041–1052. [CrossRef]
15. Amano, H. Progress and Prospect of the Growth of Wide-Band-Gap Group III Nitrides: Development of the Growth Method for Single-Crystal Bulk GaN. *Jpn. J. Appl. Phys.* **2013**, *52*, 050001. [CrossRef]
16. Amano, H.; Baines, Y.; Beam, E.; Borga, M.; Bouchet, T.; Chalker, P.R.; Charles, M.; Chen, K.J.; Chowdhury, N.; Chu, R.; et al. The 2018 GaN Power Electronics Roadmap. *J. Phys. D. Appl. Phys.* **2018**, *51*, 163001. [CrossRef]
17. Fukuda, T.; Ehrentraut, D. Prospects for the Ammonothermal Growth of Large GaN Crystal. *J. Cryst. Growth* **2007**, *305*, 304–310. [CrossRef]
18. Grabianska, K.; Kucharski, R.; Puchalski, A.; Sochacki, T.; Bockowski, M. Recent Progress in Basic Ammonothermal GaN Crystal Growth. *J. Cryst. Growth* **2020**, *547*, 125804. [CrossRef]
19. Tomida, D.; Bao, Q.; Saito, M.; Osanai, R.; Shima, K.; Kojima, K.; Ishiguro, T.; Chichibu, S.F. Ammonothermal Growth of 2 Inch Long GaN Single Crystals Using an Acidic NH4F Mineralizer in a Ag-lined Autoclave. *Appl. Phys. Express* **2020**, *13*, 055505. [CrossRef]
20. Mikawa, Y.; Ishinabe, T.; Kagamitani, Y.; Mochizuki, T.; Ikeda, H.; Iso, K.; Takahashi, T.; Kubota, K.; Enatsu, Y.; Tsukada, Y.; et al. Recent Progress of Large Size and Low Dislocation Bulk GaN Growth. In *Proceedings of SPIE, Proceedings of the Gallium Nitride Materials and Devices XV, San Francisco, CA, USA, 4–6 February 2020*; Morkoç, H., Fujioka, H., Schwarz, U.T., Eds.; SPIE-International Society for Optics and Photonics: Bellingham, WA, USA; Volume 1128002, p. 1.
21. Häusler, J.; Schnick, W. Ammonothermal Synthesis of Nitrides: Recent Developments and Future Perspectives. *Chem. A Eur. J.* **2018**, *24*, 11864–11879. [CrossRef]
22. Richter, T.; Niewa, R. Chemistry of Ammonothermal Synthesis. *Inorganics* **2014**, *2*, 29–78. [CrossRef]
23. Hertrampf, J.; Becker, P.; Widenmeyer, M.; Weidenkaff, A.; Schlücker, E.; Niewa, R. Ammonothermal Crystal Growth of Indium Nitride. *Cryst. Growth Des.* **2018**, *18*, 2365–2369. [CrossRef]
24. Mallmann, M.; Niklaus, R.; Rackl, T.; Benz, M.; Chau, T.G.; Johrendt, D.; Minár, J.; Schnick, W. Solid Solutions of Grimm-Sommerfeld Analogous Nitride Semiconductors II-IV-N2 with II = Mg, Mn, Zn; IV = Si, Ge – Ammonothermal Synthesis and DFT Calculations. *Chem. A Eur. J.* **2019**, *2*, 15887–15895. [CrossRef] [PubMed]
25. Masuda, Y.; Sato, O.; Tomida, D.; Yokoyama, C. Convection Patterns and Temperature Fields of Ammonothermal GaN Bulk Crystal Growth Process. *Jpn. J. Appl. Phys.* **2016**, *55*, 3–6. [CrossRef]
26. Erlekampf, J.; Seebeck, J.; Savva, P.; Meissner, E.; Friedrich, J.; Alt, N.S.A.; Schlücker, E.; Frey, L. Numerical Time-dependent 3D Simulation of Flow Pattern and Heat Distribution in an Ammonothermal System with Various Baffle Shapes. *J. Cryst. Growth* **2014**, *403*, 96–104. [CrossRef]
27. Jiang, Y.N.; Chen, Q.S.; Prasad, V. Numerical Simulation of Ammonothermal Growth processes of GaN Crystals. *J. Cryst. Growth* **2011**, *318*, 411–414. [CrossRef]
28. Moldovan, S.I.; Balasoiu, A.M.; Braun, M. Experimental Investigation of Natural Convection Flow in a Laterally Heated Vertical Cylindrical Enclosure. *Int. J. Heat Mass Transf.* **2019**, *139*, 205–212. [CrossRef]
29. Mirzaee, I.; Charmchi, M.; Sun, H. Heat, Mass, and Crystal Growth of GaN in the Ammonothermal Process: A Numerical Study. *Numer. Heat Transf. Part A Appl.* **2016**, *70*, 460–491. [CrossRef]
30. Pendurti, S.; Chen, Q.S.; Prasad, V. Modeling Ammonothermal Growth of GaN Single Crystals: The Role of Transport. *J. Cryst. Growth* **2006**, *296*, 150–158. [CrossRef]

31. Enayati, H.; Chandy, A.J.; Braun, M.J.; Horning, N. 3D Large Eddy Simulation (LES) Calculations and Experiments of Natural Convection in a Laterally-heated Cylindrical Enclosure for Crystal Growth. *Int. J. Therm. Sci.* **2017**, *116*, 1–21. [CrossRef]
32. Keyes, D.E.; McInnes, L.C.; Woodward, C.; Gropp, W.; Myra, E.; Pernice, M.; Bell, J.; Brown, J.; Clo, A.; Connors, J.; et al. Multiphysics Simulations: Challenges and Opportunities. *Int. J. High Perform. Comput. Appl.* **2013**, *27*, 4–83. [CrossRef]
33. John, B.; Senthilkumar, P.; Sadasivan, S. Applied and Theoretical Aspects of Conjugate Heat Transfer Analysis: A Review. *Arch. Comput. Methods Eng.* **2019**, *26*, 475–489. [CrossRef]
34. Nägel, A.; Logashenko, D.; Schroder, J.B.; Yang, U.M. Aspects of Solvers for Large-Scale Coupled Problems in Porous Media. *Transp. Porous Media* **2019**, *130*, 363–390. [CrossRef]
35. Zhang, S.; Alt, N.S.A.; Schlücker, E.; Niewa, R. Novel Alkali Metal Amidogallates as Intermediates in Ammonothermal GaN Crystal Growth. *J. Cryst. Growth* **2014**, *403*, 22–28. [CrossRef]
36. Zhang, S.; Hintze, F.; Schnick, W.; Niewa, R. Intermediates in Ammonothermal GaN Crystal Growth under Ammonoacidic Conditions. *Eur. J. Inorg. Chem.* **2013**, 5387–5399. [CrossRef]
37. Hertrampf, J.; Schlücker, E.; Gudat, D.; Niewa, R. Dissolved Intermediates in Ammonothermal Crystal Growth: Stepwise Condensation of [Ga(NH 2) 4]−toward GaN. *Cryst. Growth Des.* **2017**, *17*, 4855–4863. [CrossRef]
38. Becker, P.; Wonglakhon, T.; Zahn, D.; Gudat, D.; Niewa, R. Approaching Dissolved Species in Ammonoacidic GaN Crystal Growth: A Combined Solution NMR and Computational Study. *Chem. A Eur. J.* **2020**, *26*, 7008–7017. [CrossRef]
39. Schimmel, S.; Duchstein, P.; Steigerwald, T.G.; Kimmel, A.-C.L.; Schlücker, E.; Zahn, D.; Niewa, R.; Wellmann, P. In situ X-ray Monitoring of Transport and Chemistry of Ga-containing Intermediates under Ammonothermal Growth Conditions of GaN. *J. Cryst. Growth* **2018**, *498*, 214–223. [CrossRef]
40. Yoshida, K.; Aoki, K.; Fukuda, T. High-temperature Acidic Ammonothermal Method for GaN Crystal Growth. *J. Cryst. Growth* **2014**, *393*, 93–97. [CrossRef]
41. Grabianska, K.; Jaroszynski, P.; Sidor, A.; Bockowski, M.; Iwinska, M. GaN Single Crystalline Substrates by Ammonothermal and HVPE Methods for Electronic Devices. *Electronics* **2020**, *9*, 1342. [CrossRef]
42. Chen, Q.S.; Pendurti, S.; Prasad, V. Modeling of Ammonothermal Growth of Gallium Nitride Single Crystals. *J. Mater. Sci.* **2006**, *41*, 1409–1414. [CrossRef]
43. Schimmel, S.; Kobelt, I.; Heinlein, L.; Kimmel, A.L.; Steigerwald, T.G.; Schlücker, E.; Wellmann, P. Flow Stability, Convective Heat Transfer and Chemical Reactions in Ammonothermal Autoclaves—Insights by In Situ Measurements of Fluid Temperatures. *Crystals* **2020**, *10*, 723. [CrossRef]
44. Griffiths, S.; Pimputkar, S.; Kearns, J.; Malkowski, T.F.; Doherty, M.F.; Speck, J.S.; Nakamura, S. Growth Kinetics of Basic Ammonothermal Gallium Nitride Crystals. *J. Cryst. Growth* **2018**, *501*, 74–80. [CrossRef]
45. Malkowski, T.F.; Pimputkar, S.; Speck, J.S.; DenBaars, S.P.; Nakamura, S. Acidic Ammonothermal Growth of Gallium Nitride in a Liner-free Molybdenum Alloy Autoclave. *J. Cryst. Growth* **2016**, *456*, 21–26. [CrossRef]
46. Schimmel, S. In situ Visualisierung des Ammonothermalen Kristallisationsprozesses Mittels Röntgenmesstechnik. Ph.D. Thesis, Friedrich-Alexander-Universität (FAU), Erlangen-Nürnberg, Germany, 2018. Available online: urn:nbn:de:bvb:29-opus4-102649 (accessed on 29 March 2021).
47. Tomida, D.; Kagamitani, Y.; Bao, Q.; Hazu, K.; Sawayama, H.; Chichibu, S.F.; Yokoyama, C.; Fukuda, T.; Ishiguro, T. Enhanced Growth Rate for Ammonothermal Gallium Nitride Crystal Growth Using Ammonium Iodide Mineralizer. *J. Cryst. Growth* **2012**, *353*, 59–62. [CrossRef]
48. Letts, E.; Hashimoto, T.; Hoff, S.; Key, D.; Male, K.; Michaels, M. Development of GaN wafers via the Ammonothermal Method. *J. Cryst. Growth* **2014**, *403*, 3–6. [CrossRef]
49. Enayati, H.; Chandy, A.J.; Braun, M.J. Numerical Simulations of Transitional and Turbulent Natural Convection in Laterally Heated Cylindrical Enclosures for Crystal Growth. *Numer. Heat Transf. Part A Appl.* **2016**, *70*, 1195–1212. [CrossRef]
50. Chen, Q.S.; Pendurti, S.; Prasad, V. Effects of Baffle Design on Fluid Flow and Heat Transfer in Ammonothermal Growth of Nitrides. *J. Cryst. Growth* **2004**, *266*, 271–277. [CrossRef]
51. Masuda, Y.; Suzuki, A.; Mikawa, Y.; Kagamitani, Y.; Ishiguro, T.; Yokoyama, C.; Tsukada, T. Numerical Simulation of GaN Single-crystal Growth Process in Ammonothermal Autoclave—Effects of Baffle Shape. *Int. J. Heat Mass Transf.* **2010**, *53*, 940–943. [CrossRef]
52. Chen, Q.-S.; Jiang, Y.-N.; Yan, J.-Y.; Li, W.; Prasad, V. Modeling of Ammonothermal Growth Processes of GaN Crystal in Large-size Pressure Systems. *Res. Chem. Intermed.* **2011**, *37*, 467–477. [CrossRef]
53. Masuda, Y.; Suzuki, A.; Ishiguro, T.; Yokoyama, C. Numerical Simulation of Heat and Fluid Flow in Ammonothermal GaN Bulk Crystal Growth Process. *Jpn. J. Appl. Phys.* **2013**, *52*, 08JA05. [CrossRef]
54. Chen, Q.-S.; Prasad, V.; Hu, W.R. Modeling of Ammonothermal Growth of Nitrides. *J. Cryst. Growth* **2003**, *258*, 181–187. [CrossRef]
55. Moldovan, S.I. Numerical Simulation and Experimental Validation of Fluid Flow and Mass Transfer in an Ammonothermal Crystal Growth Reactor. Ph.D. Thesis, University of Akron, Akron, OH, USA, May 2013.
56. Schimmel, S.; Koch, M.; Macher, P.; Kimmel, A.C.L.; Steigerwald, T.G.; Alt, N.S.A.; Schlücker, E.; Wellmann, P. Solubility and Dissolution Kinetics of GaN in Supercritical Ammonia in Presence of Ammonoacidic and Ammonobasic Mineralizers. *J. Cryst. Growth* **2017**, *479*, 59–66. [CrossRef]

57. Schimmel, S.; Tomida, D.; Saito, M.; Bao, Q.; Ishiguro, T.; Honda, Y.; Chichibu, S.; Amano, H. Boundary Conditions for Simulations of Fluid Flow and Temperature Field during Ammonothermal Crystal Growth—A Machine-Learning Assisted Study of Autoclave Wall Temperature Distribution. *Crystals* **2021**, *11*, 254. [CrossRef]
58. Chen, Q.; Jiang, Y.; Yan, J.; Qin, M. Progress in modeling of Fluid Flows in Crystal Growth Processes. *Prog. Nat. Sci.* **2008**, *18*, 1465–1473. [CrossRef]
59. Li, H.; Braun, M.J.; Xing, C. Fluid Flow and Heat Transfer in a Cylindrical Model Hydrothermal Reactor. *J. Cryst. Growth* **2006**, *289*, 207–216. [CrossRef]
60. Enayati, H. Effect of Reactor Size in a Laterally-Heated Cylindrical Reactor. *Int. J. Heat Technol.* **2020**, *38*, 275–284. [CrossRef]
61. Li, H.; Evans, E.A.; Wang, G.X. Flow of Solution in Hydrothermal Autoclaves with Various Aspect Ratios. *J. Cryst. Growth* **2003**, *256*, 146–155. [CrossRef]
62. Ursu, D.; Negrila, R.; Popescu, A.; Grozescu, I.; Vizman, D. Numerical and Experimental Studies of Fluid Flow and Heat Transfer in a Model Experiment for Hydrothermal Growth. *Solid State Phenom.* **2016**, *254*, 237–242. [CrossRef]
63. Hirsch, C. *Numerical Computation of Internal and External Flows*, 2nd ed.; Elsevier: Burlington, MA, USA, 2007; ISBN 9780750665940.
64. Alfonsi, G. Reynolds-averaged Navier-Stokes Equations for Turbulence Modeling. *Appl. Mech. Rev.* **2009**, *62*, 1–20. [CrossRef]
65. Enayati, H.; Chandy, A.; Braun, M.J. Three-dimensional Large Eddy Simulations of Natural Convection in Laterally Heated Cylindrical Enclosures with Racks and Seeds for Crystal Growth. In Proceedings of the Second Thermal and Fluids Engineering Conference, Las Vegas, NV, USA, 2–5 April 2017; pp. 719–732.
66. Masuda, Y.; Suzuki, A.; Mikawa, Y.; Chani, V.; Yokoyama, C.; Tsukada, T. Numerical Simulation of Hydrothermal Autoclave for Single-Crystal Growth Process. *J. Therm. Sci. Technol.* **2008**, *3*, 540–551. [CrossRef]
67. Enayati, H.; Braun, M.J.; Chandy, A.J. Numerical Simulations of Porous Medium with Different Permeabilities and Positions in a Laterally-heated Cylindrical Enclosure for Crystal Growth. *J. Cryst. Growth* **2018**, *483*, 65–80. [CrossRef]
68. Malin, M.R. Brian Spalding: Some Contributions to Computational Fluid Dynamics During the Period 1993 to 2004. In *50 Years of CFD in Engineering Sciences*; Springer: Singapore, 2020; pp. 3–39.
69. Xiao, H.; Cinnella, P. Quantification of Model Uncertainty in RANS Simulations: A Review. *Prog. Aerosp. Sci.* **2019**, *108*, 1–31. [CrossRef]
70. Pimputkar, S.; Speck, J.S.; Nakamura, S. Basic Ammonothermal GaN Growth in Molybdenum Capsules. *J. Cryst. Growth* **2016**, *456*, 15–20. [CrossRef]
71. *Springer Handbook of Crystal Growth*; Dhanaraj, G.; Byrappa, K.; Prasad, V.; Dudley, M. (Eds.) Springer: Berlin/Heidelberg, Germany, 2010; ISBN 978-3-540-74182-4.
72. Prasad, V.; Kulacki, F.A.; Keyhani, M. Natural Convection in Porous Media. *J. Fluid Mech.* **1985**, *150*, 89–119. [CrossRef]
73. Alloui, Z.; Vasseur, P. Convection in Superposed Fluid and Porous Layers. *Acta Mech.* **2010**, *214*, 245–260. [CrossRef]
74. Battiato, I.; Ferrero V, P.T.; O' Malley, D.; Miller, C.T.; Takhar, P.S.; Valdés-Parada, F.J.; Wood, B.D. Theory and Applications of Macroscale Models in Porous Media. *Transp. Porous Media* **2019**, *130*, 5–76. [CrossRef]
75. Mao, D.; Karanikas, J.M.; Fair, P.S.; Prodan, I.D.; Wong, G.K. A Different Perspective on the Forchheimer and Ergun Equations. *SPE J.* **2016**, *21*, 1501–1507. [CrossRef]
76. Papathanasiou, T.D.; Markicevic, B.; Dendy, E.D. A Computational Evaluation of the Ergun and Forchheimer Equations for Fibrous Porous Media. *Phys. Fluids* **2001**, *13*, 2795–2804. [CrossRef]
77. Auriault, J.-L. On the Domain of Validity of Brinkman's Equation. *Transp. Porous Media* **2009**, *79*, 215–223. [CrossRef]
78. McBride, D.; Croft, T.N.; Cross, M. A Coupled Finite Volume Method for the Computational Modelling of Mould Filling in Very Complex Geometries. *Comput. Fluids* **2008**, *37*, 170–180. [CrossRef]
79. Gohil, T.; McGregor, R.H.P.; Szczerba, D.; Burckhardt, K.; Muralidhar, K.; Székely, G. Simulation of Oscillatory Flow in an Aortic Bifurcation Using FVM and FEM: A Comparative Study of Implementation Strategies. *Int. J. Numer. Methods Fluids* **2011**, *66*, 1037–1067. [CrossRef]
80. Jeong, W.; Seong, J. Comparison of Effects on Technical Variances of Computational Fluid Dynamics (CFD) Software Based on Finite Element and Finite Volume Methods. *Int. J. Mech. Sci.* **2014**, *78*, 19–26. [CrossRef]
81. Molina-Aiz, F.D.; Fatnassi, H.; Boulard, T.; Roy, J.C.; Valera, D.L. Comparison of Finite Element and Finite Volume Methods for Simulation of Natural Ventilation in Greenhouses. *Comput. Electron. Agric.* **2010**, *72*, 69–86. [CrossRef]
82. Bern, M.; Plassmann, P. Mesh Generation. In *Handbook of Computational Geometry*; North Holland: Amsterdam, The Netherlands, 2000.
83. Mirzaee, I. Computational Investigation of Gallium Nitrite Ammonothermal Crystal Growth. Ph.D. Thesis, University of Massachusetts Lowell, Lowell, MA, USA, 2015.
84. Li, H.; Evans, E.A.; Wang, G.X. A Three-dimensional Conjugate Model with Realistic Boundary Conditions for Flow and Heat Transfer in an Industry Scale Hydrothermal Autoclave. *Int. J. Heat Mass Transf.* **2005**, *48*, 5166–5178. [CrossRef]
85. Zajac, M.; Kucharski, R.; Grabiańska, K.; Gwardys-Bąk, A.; Puchalski, A.; Boćkowski, M. Ammonothermal GaN Substrates for Microwave Electronics and Energoelectronics. In *Proceedings of SPIE, Proceedings of the Radioelectronics Systems Conference, Jachranka, Poland, 14–16 November 2017*; SPIE-International Society for Optics and Photonics: Bellingham, WA, USA. [CrossRef]
86. Pimputkar, S.; Kawabata, S.; Speck, J.S.; Nakamura, S. Surface Morphology Study of Basic Ammonothermal GaN Grown on Non-polar GaN Seed Crystals of Varying Surface Orientations from m-plane to a-plane. *J. Cryst. Growth* **2013**, *368*, 67–71. [CrossRef]

87. Alt, N.; Meissner, E.; Schlücker, E.; Frey, L. In situ Monitoring Technologies for Ammonthermal Reactors. *Phys. Status Solidi Curr. Top. Solid State Phys.* **2012**, *9*, 436–439. [CrossRef]
88. Steigerwald, T.G.; Balouschek, J.; Hertweck, B.; Kimmel, A.-C.L.; Alt, N.S.A.; Schluecker, E. In situ Investigation of Decomposing Ammonia and Ammonobasic Solutions under Supercritical Conditions via UV/vis and Raman Spectroscopy. *J. Supercrit. Fluids* **2018**, *134*, 96–105. [CrossRef]
89. Schimmel, S.; Lindner, M.; Steigerwald, T.G.; Hertweck, B.; Richter, T.M.M.; Künecke, U.; Alt, N.S.A.; Niewa, R.; Schlücker, E.; Wellmann, P.J. Determination of GaN Solubility in Supercritical Ammonia with NH4F and NH4Cl Mineralizer by in situ X-ray Imaging of Crystal Dissolution. *J. Cryst. Growth* **2015**, *418*, 64–69. [CrossRef]
90. Schimmel, S.; Künecke, U.; Meisel, M.; Hertweck, B.; Steigerwald, T.G.; Nebel, C.; Alt, N.S.A.; Schlücker, E.; Wellmann, P. Chemical Stability of Carbon-based Inorganic Materials for in situ X-ray Investigations of Ammonothermal Crystal Growth of Nitrides. *J. Cryst. Growth* **2016**, *456*, 33–42. [CrossRef]
91. Schimmel, S.; Wellmann, P. In situ Visualization of the Ammonothermal Crystallization Process by X-ray Technology. In *Ammonothermal Synthesis and Crystal Growth of Nitrides–Chemistry and Technology*; Niewa, R., Meissner, E., Eds.; Springer International Publishing: Berlin, Germany, 2021; Volume 304, ISBN 978-3-030-56305-9.
92. Beckermann, C.; Viskanta, R.; Ramadhyani, S. Natural Convection in Vertical Enclosures Containing Simultaneously Fluid and Porous Layers. *J. Fluid Mech.* **1988**, *186*, 257–284. [CrossRef]
93. Wang, B.; Callahan, M.J.; Rakes, K.D.; Bouthillette, L.O.; Wang, S.Q.; Bliss, D.F.; Kolis, J.W. Ammonothermal Growth of GaN Crystals in Alkaline Solutions. *J. Cryst. Growth* **2006**, *287*, 376–380. [CrossRef]
94. Griffiths, S.; Pimputkar, S.; Speck, J.S.; Nakamura, S. On the Solubility of Gallium Nitride in Supercritical Ammonia–sodium Solutions. *J. Cryst. Growth* **2016**, *456*, 5–14. [CrossRef]
95. Mirzaii, I.; Passandideh-Fard, M. Modeling Free Surface Flows in Presence of an Arbitrary Moving Object. *Int. J. Multiph. Flow* **2012**, *39*, 216–226. [CrossRef]
96. Liu, X.; Harada, H.; Miyamura, Y.; Han, X.F.; Nakano, S.; Nishizawa, S.I.; Kakimoto, K. Transient Global Modeling for the Pulling Process of Czochralski Silicon Crystal Growth. I. Principles, Formulation, and Implementation of the Model. *J. Cryst. Growth* **2020**, *532*, 125405. [CrossRef]
97. Dadzis, K.; Pätzold, O.; Gerbeth, G. Model Experiments for Flow Phenomena in Crystal Growth. *Cryst. Res. Technol.* **2020**, *55*, 1–7. [CrossRef]
98. Lemmon, E.W.; McLinden, M.O.; Friend, D.G. *Thermophysical Properties of Fluid Systems*; Linstrom, P.J., Mallard, W.G., Eds.; Available online: http://http//webbook.nist.gov/chemistry (accessed on 5 June 2017).
99. Tomida, D.; Yoshinaga, T. Thermal Conductivity Measurements of Liquid Ammonia by the Transient Short-Hot-Wire Method. *Int. J. Thermophys.* **2020**, *41*, 53. [CrossRef]
100. Brennecke, J.F.; Eckert, C.A. Phase Equilibria for Supercritical Fluid Process Design. *AIChE J.* **1989**, *35*, 1409–1427. [CrossRef]
101. *Supercritical Fluids*; Arai, Y.; Sako, T.; Takebayashi, Y. (Eds.) Springer: Berlin/Heidelberg, Germany, 2002; ISBN 978-3-642-62515-2.
102. Pimputkar, S.; Nakamura, S. Decomposition of Supercritical Ammonia and Modeling of Supercritical Ammonia-nitrogen-hydrogen Solutions with Applicability toward Ammonothermal Conditions. *J. Supercrit. Fluids* **2016**, *107*, 17–30. [CrossRef]
103. Hosaka, M.; Taki, S. Hydrothermal Growth of Quartz Crystals in Pure Water. *J. Cryst. Growth* **1981**, *51*, 640–642. [CrossRef]
104. Hervey, P.R.; Foise, J.W. Syntheic Quartz Crystal—A Review. *Miner. Metall. Process.* **2001**, *18*, 1–4.
105. Ehrentraut, D.; Sato, H.; Kagamitani, Y.; Sato, H.; Yoshikawa, A.; Fukuda, T. Solvothermal Growth of ZnO. *Prog. Cryst. Growth Charact. Mater.* **2006**, *52*, 280–335. [CrossRef]
106. Bao, Q.; Saito, M.; Hazu, K.; Kagamitani, Y.; Kurimoto, K.; Tomida, D.; Qiao, K.; Ishiguro, T.; Yokoyama, C.; Chichibu, S.F. Ammonothermal Growth of GaN on a Self-nucleated GaN Seed Crystal. *J. Cryst. Growth* **2014**, *404*, 168–171. [CrossRef]
107. Duan, X.; Qian, G.; Fan, C.; Zhu, Y.; Zhou, X.; Chen, D.; Yuan, W. First-principles Calculations of Ammonia Decomposition on Ni(110) Surface. *Surf. Sci.* **2012**, *606*, 549–553. [CrossRef]
108. D'Evelyn, M.P.; Ehrentraut, D.; Jiang, W.; Kamber, D.S.; Downey, B.C.; Pakalapati, R.T.; Yoo, H.D. Ammonothermal Bulk GaN Substrates for Power Electronics. *ECS Trans.* **2013**, *58*, 287–294. [CrossRef]
109. Hertweck, B.; Schimmel, S.; Steigerwald, T.G.; Alt, N.S.A.; Wellmann, P.J.; Schluecker, E. Ceramic Liner Technology for Ammonoacidic Synthesis. *J. Supercrit. Fluids* **2015**, *99*, 76–87. [CrossRef]
110. Bajus, S. Ammoniakzersetzung mit Salzmodifizierten Katalysatoren. Ph.D. Thesis, Friedrich-Alexander-Universität Erlangen-Nürnberg (FAU), Erlangen-Nürnberg, Germany, June 2014.
111. Lu, K.; Tatarchuk, B.J. Activated Chemisorption of Hydrogen on Supported Ruthenium. II. Effects of Crystallite Size and Adsorbed Chlorine on Accurate Surface Area Measurements. *J. Catal.* **1987**, *106*, 176–187. [CrossRef]
112. Shlflett, W.K.; Dumeslc, J.A. Ammonia Synthesis as a Catalytic Probe of Supported Ruthenium Catalysts: The Role of the Support and the Effect of Chlorine. *Ind. Eng. Chem. Fundam.* **1981**, *20*, 246–250. [CrossRef]
113. Steigerwald, T.G.; Alt, N.S.A.; Hertweck, B.; Schluecker, E. Feasibility of Density and Viscosity Measurements under Ammonothermal Conditions. *J. Cryst. Growth* **2014**, *403*, 59–65. [CrossRef]
114. Keblinski, P.; Eastman, J.A.; Cahill, D.G. Nanofluids for Thermal Transport. *Mater. Today* **2005**, *8*, 36–44. [CrossRef]
115. Buongiorno, J. Convective Transport in Nanofluids. *J. Heat Transfer* **2006**, *128*, 240–250. [CrossRef]
116. Bänsch, E. A Thermodynamically Consistent Model for Convective Transport in Nanofluids: Existence of Weak Solutions and Fem Computations. *J. Math. Anal. Appl.* **2019**, *477*, 41–59. [CrossRef]

117. Tomida, D.; Kuroda, K.; Nakamura, K.; Qiao, K.; Yokoyama, C. Temperature Dependent Control of the Solubility of Gallium Nitride in Supercritical Ammonia Using Mixed Mineralizer. *Chem. Cent. J.* **2018**, *12*, 1–6. [CrossRef] [PubMed]
118. Tomida, D.; Kuroda, K.; Hoshino, N.; Suzuki, K.; Kagamitani, Y.; Ishiguro, T.; Fukuda, T.; Yokoyama, C. Solubility of GaN in Supercritical Ammonia with Ammonium Chloride as a Mineralizer. *J. Cryst. Growth* **2010**, *312*, 3161–3164. [CrossRef]
119. Bushmin, S.A.; Azimov, P.; Lvov, S. Numerical Modelling of the Metamorphic Mineral Solubility in Hydrothermal Solutions at 400-800 C, 1-5 kbar and Various Fluid Acidity. *Mineral. Collect.* **2004**, *54*, 94–116.
120. Zahn, D. On the Solvation of Metal Ions in Liquid Ammonia: A Molecular Simulation Study of M(NH2)x(NH3)y Complexes as a Function of pH. *RSC Adv.* **2017**, *7*, 54063–54067. [CrossRef]
121. Zahn, D. A Molecular Simulation Study of the Auto-protolysis of Ammonia as a Function of Temperature. *Chem. Phys. Lett.* **2017**, *682*, 55–59. [CrossRef]
122. Ehrentraut, D.; Kagamitani, Y.; Yokoyama, C.; Fukuda, T. Physico-chemical Features of the Acid Ammonothermal Growth of GaN. *J. Cryst. Growth* **2008**, *310*, 891–895. [CrossRef]
123. Hashimoto, T.; Saito, M.; Fujito, K.; Wu, F.; Speck, J.S.; Nakamura, S. Seeded Growth of GaN by the Basic Ammonothermal Method. *J. Cryst. Growth* **2007**, *305*, 311–316. [CrossRef]
124. Hashimoto, T.; Letts, E. Development of Cost-Effective Native Substrates for Gallium Nitride-Based Optoelectronic Devices via Ammonothermal Growth. *Optoelectron. Devices Appl.* **2012**, 95–106. [CrossRef]
125. Schimmel, S.; Künecke, U.; Baser, H.; Steigerwald, T.G.; Hertweck, B.; Alt, N.S.A.; Schlücker, E.; Schwieger, W.; Wellmann, P. Towards X-ray in-situ Visualization of Ammonothermal Crystal Growth of Nitrides. *Phys. Status Solidi Curr. Top. Solid State Phys.* **2014**, *11*, 1439–1442. [CrossRef]
126. Font, B.; Weymouth, G.D.; Nguyen, V.-T.; Tutty, O.R. Deep Learning of the Spanwise-averaged Navier–Stokes Equations. *J. Comput. Phys.* **2021**, *434*, 110199. [CrossRef]
127. Tsunooka, Y.; Kokubo, N.; Hatasa, G.; Harada, S.; Tagawa, M.; Ujihara, T. High-speed Prediction of Computational Fluid Dynamics Simulation in Crystal Growth. *CrystEngComm* **2018**, *20*, 6546–6550. [CrossRef]
128. Raissi, M.; Perdikaris, P.; Karniadakis, G.E. Physics-informed Neural Networks: A Deep Learning Framework for Solving Forward and Inverse Problems Involving Nonlinear Partial Differential Equations. *J. Comput. Phys.* **2019**, *378*, 686–707. [CrossRef]
129. Givi, P.; Daley, A.J.; Mavriplis, D.; Malik, M. Quantum Speedup for Aeroscience and Engineering. *AIAA J.* **2020**, *58*, 3715–3727. [CrossRef]
130. Bharadwaj, S.S.; Sreenivasan, K.R. Quantum Computation of Fluid Dynamics. 2020. Available online: https://arxiv.org/abs/2007.09147 (accessed on 29 March 2021).
131. Bockowski, M.; Iwinska, M.; Amilusik, M.; Fijalkowski, M.; Lucznik, B.; Sochacki, T. Challenges and Future Perspectives in HVPE-GaN Growth on Ammonothermal GaN Seeds. *Semicond. Sci. Technol.* **2016**, *31*, 093002. [CrossRef]
132. Gao, B.; Kakimoto, K. Three-dimensional Modeling of Basal Plane Dislocations in 4H-SiC Single Crystals Grown by the Physical Vapor Transport Method. *Cryst. Growth Des.* **2014**, *14*, 1272–1278. [CrossRef]

Review

Current Understanding of Bias-Temperature Instabilities in GaN MIS Transistors for Power Switching Applications

Milan Ťapajna

Institute of Electrical Engineering, Slovak Academy of Sciences, Dúbravská cesta 9, 84104 Bratislava, Slovakia; milan.tapajna@savba.sk; Tel.: +421-2-5922-2777

Received: 30 November 2020; Accepted: 14 December 2020; Published: 18 December 2020

Abstract: GaN-based high-electron mobility transistors (HEMTs) have brought unprecedented performance in terms of power, frequency, and efficiency. Application of metal-insulator-semiconductor (MIS) gate structure has enabled further development of these devices by improving the gate leakage characteristics, gate controllability, and stability, and offered several approaches to achieve E-mode operation desired for switching devices. Yet, bias-temperature instabilities (BTI) in GaN MIS transistors represent one of the major concerns. This paper reviews BTI in D- and E-mode GaN MISHEMTs and fully recess-gate E-mode devices (MISFETs). Special attention is given to discussion of existing models describing the defects distribution in the GaN-based MIS gate structures as well as related trapping mechanisms responsible for threshold voltage instabilities. Selected technological approaches for improving the dielectric/III-N interfaces and techniques for BTI investigation in GaN MISHEMTs and MISFETs are also outlined.

Keywords: GaN transistors; MIS/MOS; MISHEMT; MISFET; PBTI; NBTI; threshold voltage instability; interface traps; oxide traps

1. Introduction

A combination of wide band gap (3.4 eV), high breakdown electric field (3 MV/cm), decent thermal conductivity (>1.5 W/cmK), and high saturation velocity (~10^7 cm/s) of electrons makes GaN an ideal material for high-power semiconductor devices [1–3]. Indeed, GaN-based high electron-mobility transistors (HEMTs) with high cut-off frequency and high breakdown voltage (V_{BD}) enabled development of new generation of power amplifiers implemented in wireless communication, satellite, and radar systems commercially available already a decade ago [4]. More recently, GaN HEMTs have been also applied as switching devices for power converters. Despite relatively immature technology, the state-of-the-art GaN switching devices have shown lower ON-state resistance (R_{ON}) for given V_{BD} compared to current power devices based on Si [5–7]. Intensive R&D effort in the last decade pawed the way to emerging of highly efficient and compact GaN-based power converters in the market [8]. However, the issues related to stability and reliability of GaN power switching devices hamper a more dramatic commercial success of this technology. To take advantage of outstanding properties of GaN material, a key task is to gain a fundamental understanding of the parasitic and degradation mechanisms that negatively affect the performance and long-term reliable operation of these devices. This represents a rather difficult task, keeping in mind the unique properties of GaN-based materials (wide-band gap nature, piezoelectricity) and high electric field combined with dissipating power of the operating devices. In addition, GaN heterostructures for lateral transistors are grown on foreign substrates. Therefore, a variety of extended defects are present in the device active region.

GaN HEMTs with Schottky-barrier gates often suffer from relatively large gate leakage current (I_G) [9]. An effective way to suppress the excessive I_G is to employ metal-insulator-semiconductor (MIS) gate structure. MISHEMTs with largely suppressed I_G as compared to Schottky-gated devices has been reported by many groups using various gate dielectrics including Al_2O_3, Si_3N_4, SiO_2, AlN, HfO_2, and others [9]. By suppressing I_G in particular at forward bias, MISHEMTs with improved gate controllability and stability under DC as well as RF operation have been reported [10]. In addition, MIS gate structure offer several approaches to achieve E-mode transistor operation, which is highly desired for switching devices. Proposed E-mode concepts are based on increased gate capacitance using partial [11] or full barrier recessing [12] and/or introduction of sufficiently high negative charge at the dielectric/barrier interface [13] or in the dielectric layer itself [14]. Yet, dielectric/barrier interface in the MIS gate structure inevitably contains interfacial defect levels that can interact with free electrons from the 2DEG channel or metal electrode [15–18]. Depending on the distribution of these traps, MISHEMTs suffer from threshold voltage (V_{TH}) stability issues, known as bias temperature instability (BTI) in the literature.

BTI represents a reliability issue, manifested by the change of the transistor's V_{TH} under applied gate bias, resulting in the change of the drain current (I_D) and transconductance (g_m) of the device. It is generally enhanced by the stress voltage and temperature. BTI effects can be recoverable at less severe conditions and originate from the trapping effects. In the harsher conditions, new traps in the gate stack (typically interface states) can be also formed [19], leading to permanent change of V_{TH}. In BTI test, either positive (PBTI) or negative (NBTI) bias is applied on the gate electrode (while the source and drain electrodes are grounded, i.e., $V_{DS} = 0$) at elevated temperature and change of the electrical parameters during both, stressing and recovery period is monitored.

In GaN MISHEMTs, BTI represents one of the biggest reliability concerns. This is due to relatively high density of traps located in the gate stack, being a consequence of unavailability of high-quality native oxides for GaN-based semiconductors and complexity of the dielectric/III-N interfaces. As a result, considerable BTI with V_{TH} instabilities ranging from 100 mV up to several V have been commonly reported for GaN MISHEMTs in the literature [20,21]. For D-mode MISHEMTs, NBTI is expected to be a major concern as the device is commonly biased in OFF-state with $V_{GS} < 0$. Indeed, several researchers have reported NBTI to induce negative V_{TH} drift strongly enhanced by temperature [21]. Although PBTI may be considered less problematic in these devices, many studies have been devoted to PBTI investigations in D-mode MISHEMTs with an aim to analyze the underlying mechanism of V_{TH} drift [22]. The dynamics of the PBTI in GaN MISHEMTs was found to differ from that known for Si metal-oxide-semiconductor field-effect transistors (MOSFET), mostly because of coaction of different trapping states in the gate stack, nontrivial defect dynamics, and electron transport over the existing barrier affecting the trapping dynamic [23]. For E-mode MISHEMTs, PBTI is clearly the major concern as the positive V_{GS} drives the device into ON-state. Dramatic V_{TH} drifts upon PBTI testing has been reported in the literature [24]. In addition, specific designs for achieving E-mode behavior for MISHEMTs, such as application of InGaN/AlGaN double barrier layer, have been shown to result in a unique mechanism of PBTI [25]. GaN transistors with fully recessed barrier (also known as recessed gate hybrid MISHEMTs [26]) represent a special design of E-mode GaN devices, refereed here to as GaN MISFETs. A complete etching-away of the barrier layer under the gate greatly simplifies the interpretation of the BTI data. Available studies investigating BTI in GaN MISFETs consistently indicate that both PBTI and NBTI need to be concerned. Moreover, it seems that dielectric bulk traps with specific distribution play a major role affecting the PBTI as well as NBTI behavior [27,28].

Up to now, BTI has been reviewed separately for D-mode MISHEMTs [21,23] and E-mode GaN MISFETs [29]. Intention of this review is to provide a full picture of BTI phenomenon in GaN MIS-gated switching transistors, based on most recent reports on D- and E-mode GaN MISHEMTs as well as E-mode GaN MISFETs. In particular, we focus on the existing models for origin of defect states present in the GaN MIS gate stacks and the underlaying physics of their capture and emission processes leading to V_{TH} drift. The paper is organized as follows: In Section 2, we will describe the device concepts, existing models for trap distribution in different MIS gate structures, related trapping mechanisms, and the methods used for BTI investigation. In Sections 3 and 4, we will review most recent studies of

BTI in GaN MISHEMTs and MISFETs, respectively. Finally, the summary and prospects will be given in Section 5.

2. Devices, Models, and Methods

2.1. GaN MISHEMTs and MISFETs

The schematic of a lateral GaN MISHEMT is depicted in Figure 1a. The device concept employs the effect of polarization charges at the heterointerface between the GaN channel and a thin barrier layer (AlGaN, AlInN, AlN), originating from the difference between spontaneous and/or piezoelectric polarization of these layers. This essentially fixed charge gives rise to formation of 2-dimentional electron gas (2DEG) with a high density of free electrons (~0.8–3 × 10^{13} cm^{-2}) in the GaN channel [30]. In contrast to Schottky barrier gate, the insulated gate ensures the suppression of the excessive gate leakage and extends the gate voltage span towards the positive values. Gate dielectric/barrier/GaN MIS structure includes two interfaces that in general contain fixed polarization charge as well as defect states, which can exchange their charge with the carrier reservoirs in the 2DEG and metal electrode.

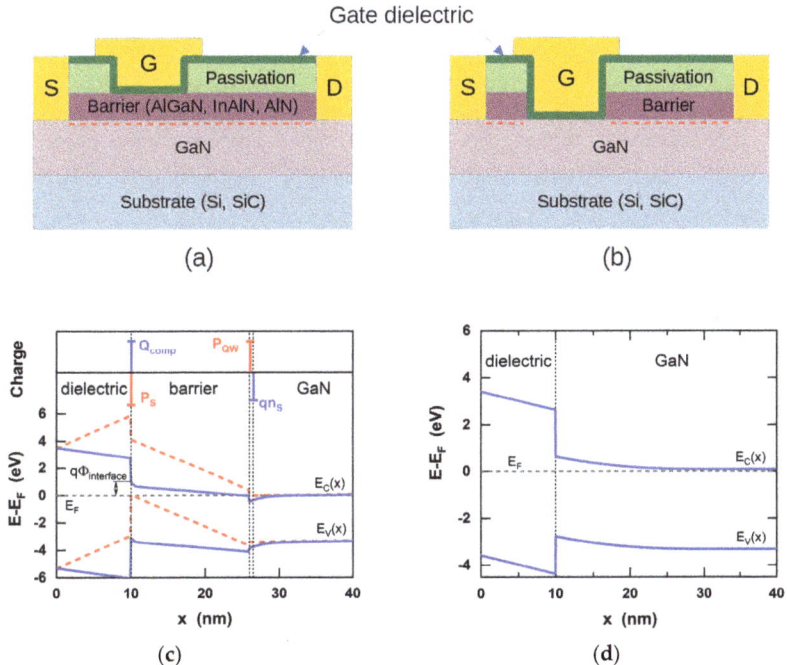

Figure 1. The schematic of the lateral GaN metal-insulator-semiconductor high-electron mobility transistors (MISHEMT) (**a**) and fully barrier recess-gate MIS field effect transistor (MISFET) (**b**) with corresponding band-diagrams across the gate structure (**c**,**d**). In (**c**), interface potential $\Phi_{interface}$ is defined and also charge distributions with fully compensated (solid blue lines) and uncompensated (dashed red lines) surface polarization charge P_S are depicted.

The band diagram of MISHEMT across the gate structure is shown in Figure 1c for heterostructure with Ga-face polarity [30]. Note that the electrically neutral dielectric/barrier interface is implicitly assumed in this model. This means that the negative polarization charge located at the surface of the barrier (P_S) is assumed to be fully compensated by some charge of similar density and opposite polarity (Q_{comp}), i.e., $P_S + Q_{comp} \approx 0$. Otherwise, uncompensated P_S would rise the bands at the interface and deplete the 2DEG channel (shown by the dashed lines in Figure 1c), which is in clear contrast to commonly observed behavior of GaN MISHEMTs. The origin and nature of the compensating charge is not fully clear yet and remains the subject of debate. It is commonly accepted that P_S is compensated by surface donor states with a single level [31] or distribution of levels [32] with the density well above 10^{13} cm^{-2}, also providing free electrons into 2DEG, as originally proposed by Ibbetson et al. [31]. In MISHEMT structure, these surface donor states should act as interface states and contribute to PBTI [33]. Alternatively, acceptor-like interface states causing PBTI may coexist with the donor states defining the 2DEG density [23]. However, the lack of correlation between evaluated interface state density and P_S observed by several researchers [34–37] led to introduction of other models for origin of Q_{comp}. These models assume fixed charge formed by the ionized donor states between conduction band (CB) edges of the barrier and the dielectric layer [35,38,39], or the near-interface traps located spatially in the dielectric and energetically below the barrier CB edge [34]. Recently, Ber et al. [37] proposed a surface polarization self-compensation mechanism. The authors speculated that in contrast to ordered arrangement of ions at the epitaxial barrier/GaN interface, displacement of less rigid surface ions may be responsible for P_S self-compensation effect [37].

2.1.1. E-Mode GaN MIS Transistors

Due to the equilibrium population of 2DEG in the GaN channel, GaN MISHEMTs are inherently D-mode devices. Considering the converter topology and its safe operation, it is strongly desirable to implement the switching devices with E-mode transistors with sufficiently high positive V_{TH} [40]. This requirement ensures the robustness of the switch against the unwanted random turn-on and avoid hazardous voltage on the terminals in the case of control electronics failure. There are two fundamental approaches for achieving E-mode operation of GaN MISHEMTs: (i) introduction of sufficiently high (density $> 1.8 \times 10^{13}$ cm^{-2} [41]) of negative interface charge ($Q_{interface}$) leading to increase of the interface potential ($\Phi_{interface}$, see Figure 1c) and (ii) increasing of the gate capacitance, preferably by decreasing of the barrier thickness [42]. Processing of the E-mode MISHEMTs using the first approach have been reported by employing F plasma treatment (fluorination) underneath the gate [14], control over the $Q_{interface}$ [43–46], and polarization engineering concept using InGaN/AlGaN barrier double-layer [13].

Using the second approach, preparation of E-mode GaN MISHEMTs with partially recess-barrier under the gate have been reported [47,48]. In the limiting case, the barrier under the gate can be fully recessed (Figure 1b), which provides E-mode transistor operation [12,48–51]. The band structure exemplified in Figure 1d shows that the electrons under the gate are depleted and interrupt the conductive channel between the source and drain access regions (Figure 1b). As this device connects in series the recessed MIS channel with two access regions having a low resistance due to the presence of 2DEG, it is often referred to as GaN recess-gate hybrid MISHEMTs [49,50] or MISFETs [12,52] in the literature. In this review, we will refer to these devices as GaN MISFETs. In contrast to inversion type Si MOSFETs, GaN MISFETs are majority carrier transistors. R_{ON} is given by relatively low access resistances and the intrinsic channel resistance, proportional to the gate length and the GaN channel mobility, which is a strong function of the dielectric/GaN interface quality.

2.1.2. Gate Materials and Technologies

The main requirements of an MIS gate structure are suppression of the gate leakage even at forward bias and its stability at different operating conditions. In addition, the process technology needs to be robust and limits the resulting V_{TH} dispersion. For an optimal design of the gate structure, it is necessary to consider the bandgap, band offsets, dielectric constant, breakdown field, and chemical stability of the dielectric with III-N semiconductor. For suppression of the gate leakage current, a dielectric with large band offsets in respect to GaN needs to be selected. Although the dielectrics with high permittivity (high-k dielectrics) allow fabrication of transistors with high g_m, the general tradeoff between bandgap and dielectric constant restricts the applicability of several high-k dielectrics for GaN MISHEMTs [9]. Importantly, the defects in the dielectric and its interface with semiconductor affects the V_{TH} stability and, in the case of GaN MISFETs, also channel mobility [26].

In addition to the selection of the gate dielectric, it is as well necessary to adopt a suitable technology including the III-N surface pre-treatment, gate dielectric deposition method, and MIS gate post-treatment. Several high-k dielectrics such as HfO_2, ZrO_2, $GdScO_3$, Ta_2O_5, $LaLuO_3$, La_2O_3, $MgCaO$, TiO_2 [53–59], etc. have been applied as the gate dielectrics in GaN MISHEMTs. However, nowadays, Al_2O_3 (and Al-based oxides), SiO_2, and SiN_x are most commonly used dielectrics materials. Al_2O_3 has relatively large band gap (6.7–7.0 [60]), high breakdown filed (~10 MV/cm) and sufficiently large band off-stets in respect to AlGaN/GaN. It is typically grown by atomic layer deposition (ALD) using various oxidation agents (H_2O [15,61], O_2-plasma [62], O_3 [63]) at moderate temperatures ranging from 100 to 300 °C. High-quality ALD Al_2O_3 gate dielectrics for GaN MISHEMTs and MISFETs have been reported by several groups, exhibiting low gate leakage current, nearly theoretical breakdown field, and excellent interface properties [61,64–67]. Growth of Al_2O_3 using MOCVD at higher temperatures (600 °C) has been also reported [55]. In order to improve the thermal stability of Al_2O_3 beyond 800 °C, application of aluminum oxynitride (AlON) and Al_2O_3/SiO_2 nanolaminates grown by ALD has been demonstrated [68,69].

Despite its lower dielectric constant (3.9), SiO_2 has the highest band gap (~9 eV) among other dielectrics. It is typically deposited by plasma-enhanced chemical vapor deposition (PECVD) with subsequent post-deposition annealing (PDA). High-quality SiO_2 gate dielectrics with breakdown field reaching 11 MV/cm [70] have been applied to GaN MISHEMTs [71] and MISFETs [72] with excellent electrical properties. Although SiN_x has been primarily used as the passivation layer in GaN HEMTs, several groups reported its application as the gate dielectric [27,73–75]. SiN_x can be grown by in-situ MOCVD, PECVD [73,74], low-pressure CVD, and plasma-enhanced ALD [27,75]. Although MISFETs with SiN_x exhibit good dielectric/GaN interface quality [73], MISHEMTs often show relatively high gate leakage at forward bias due to small CB offset [76]. This issue has been overcome by e.g., deposition of the SiN_x/Al_2O_3 dielectric bilayer [77]. Finally, the quality of dielectric-III/N interface can be improved by surface pre-treatment such as Cl_2-based inductively coupled plasma etching [61], in-situ remote plasma treatment [78], application of AlN interlayer [79], or the gate metal post-treatment [80].

2.2. Modeling of Defect States in the Gate Stack

BTI is dominantly affected by the dynamics of various trapping states in the gate stack. In the following, we will summarize current understanding of the defect's origin, their nature and spatial distribution in GaN MIS gate structure. We will also describe the most relevant empirical models of the defect states and the observed BTI behavior proposed in the literature.

2.2.1. Interface Traps

Interface traps (IT) represent allowed states in the semiconductor bandgap located at its interface with the dielectric, as depicted in Figure 2a. In general, IT can be divided into intrinsic states of the semiconductor surface and extrinsic interface defects. Intrinsic interface states are associated with the surface reconstruction of the crystal termination and, for III-N surfaces, they are most likely formed by the vacancies and dangling bonds [81]. Extrinsic interface defects originate from adsorbed foreign atoms, sub-oxides, and structural imperfections and in general depends upon semiconductor surface cleaning, dielectric material, and the method of its deposition. Although several IT models have been proposed in the literature (for more comprehensive review see, e.g., Eller et al. [82]), no specific IT model exist for dielectric/III-N interfaces. Yet, several researchers [16,76,83] adopted disorder-induced gap state (DIGS) model proposed by Hasegawa et al. [84], originally introduced for III-V MIS structures. The model assumes existence of several monolayers thick disordered layer at the semiconductor surface with distortion of lengths and angles of the local bonds. This leads to formation of continuum of IT states energetically distributed with typical U-shape within the semiconductor bandgap $D_{IT}(E_{IT})$ (depicted in Figure 2b). Here, acceptor-like (anti-bonding) trap states and donor-like (bonding) trap states are divided by charge neutrality level (E_{CNL}) [84]. Although it is challenging to experimentally distinguish between response of IT and other traps in the gate stack, relatively high D_{IT} in the range of 10^{12}–10^{13} eV^{-1}cm^{-2} has been consistently reported for GaN MISHEMT structures using various methods, including capacitance–voltage (CV) method [17,85], photo-assisted CV [18], C-transients [16], deep-level transient spectroscopy (DLTS) [86], and AC admittance (C-ω, G-ω) techniques [15,78]. However, special care must be taken in the interpretation of these techniques for D_{IT} evaluation as the gate bias and temperature dependence of the barrier conductivity can affect the frequency response of the MISHEMT gate admittance even without the presence of IT [87].

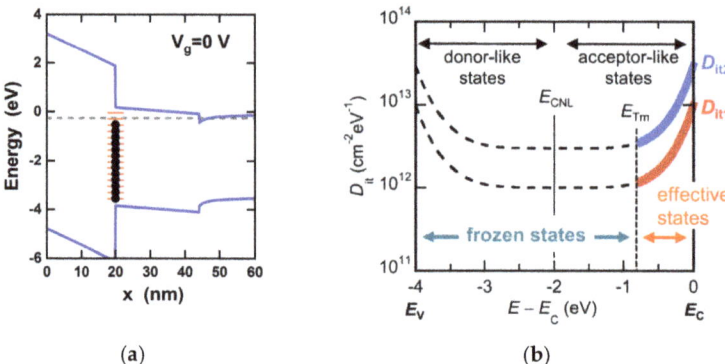

Figure 2. (a) Schematic illustration of interface traps in Al$_2$O$_3$/AlGaN/GaN MIS heterostructure. (b) Example of two U-shape D_{IT} energy distributions at Al$_2$O$_3$/AlGaN interface used for capacitance–voltage (CV) simulations, highlighting interface traps in the upper part of the bandgap that affect the CV stretch-out. Panel (b) reprinted from [15], with the permission of AIP Publishing. Copyright © 2020 AIP.

A wide band gap nature of GaN based semiconductors needs to be also considered in evaluation of D_{IT} energy distribution in GaN MIS heterostructures. First, due to extremely low density of holes in GaN, hole capture/emission can be readily neglected. Second, IT with a wide range of energies in the semiconductor band gap are characterized by extremally wide range of time responses. Using classical Shockley–Read–Hall (SRH) theory, capture and emission processes can be described in terms of capture and emission time constants [88]

$$\tau_C = \frac{1}{n v_{TH} \sigma}, \qquad (1)$$

$$\tau_E = \frac{1}{N_C v_{TH} \sigma} \exp\left(\frac{E_C - E_{IT}}{k_B T}\right) \qquad (2)$$

where n, N_C, v_{TH}, σ, E_C-E_{IT}, k_B, and T are the concentration of electrons and effective density of states at the conduction band, the thermal velocity of electrons, the capture cross-section and IT energy position measured from the CB bottom, the Boltzmann constant, and temperature, respectively. While τ_C depends only on σ and the number of free electrons available for capture, τ_E also depends exponentially on E_{IT}, because trapped electron must gain enough thermal energy to be transferred to CB. As pointed out by Miczek et al. [88], only shallower IT with $E_C - E$ < ~1 eV (assuming $\sigma = 10^{-16}$ cm^2 and room temperature) are capable to emit electrons into conduction band of the semiconductor within 100 s, i.e., practical time for CV measurements. For IT with $E_C - E$ > ~1 eV, however, V_G-induced Fermi level movement below the trap level towards the valence band (VB) does not change their occupation and these traps remain frozen, as depicted in Figure 2b [15]. This implies a fundamental limitation for application of standard capacitance and admittance methods used for $D_{IT}(E_{IT})$ determination.

IT represents one of the major concerns in relation to BTI in GaN MISHEMTs. In fact, relatively large variation in reported D_{IT} ranging from 10^{11} up to 10^{13} eV^{-2}cm^{-2} can be found in the literature. Yet, recently, Hashizume et al. [80] has reported ALD-grown Al$_2$O$_3$/GaN-on-GaN MIS structures with superior interface quality, as documented by D_{IT} on the level of 10^{10} eV^{-2}cm^{-2}. This illustrates that combination of high-quality GaN channel region, optimized ALD process, careful surface pre-treatment, and post-deposition annealing can provide dielectric/GaN interface with quality similar to that of SiO$_2$/Si interface.

2.2.2. Disorder-Induced Gap States in the Gate Dielectric

Existence of DIGS in the gate dielectric has been proposed by Matys et al. [89] in order to describe complex PBTI and NBTI behavior of Al$_2$O$_3$/AlGaN/GaN MISHEMT structures. Similar to DIGS model discussed above, U-shaped energy distribution of DIGS that exponentially decay toward the dielectric bulk (up to 4 nm) from its interface with the barrier is assumed in this model. The density of DIGS, their energy, and spatial distribution depend upon nature and degree of the disorder introduced upon the gate dielectric technology. It can be expressed as [90]

$$N_{DIGS}(E,x) = N_0 exp\left(\left|\frac{E - E_{CNL}}{E_{0d,0a}}\right|^{n_{d,a}}\right) exp\left(-\frac{x}{x_l}\right) \qquad (3)$$

where N_0 is the DIGS minimum density, E_{0d}/E_{0a} and n_d/n_a describe the energy shape of the donor/acceptor-like branch, respectively, and x_l describes the DIGS spatial distribution. The capture and emission behavior of DIGS is then described by SRH statistics and tunneling-assisted processes with x-dependent capture cross-section expressed as [91]

$$\sigma(x) = \sigma_0 exp(-x/x_0). \qquad (4)$$

In Equation (4), σ_0 is the electron capture cross-section of the states at the interface and x_0 is the tunneling decay length given by $x_0 = \hbar/\sqrt{2m_e \Delta E_C}$, where m_e is the effective mass of the electron and ΔE_C is the CB offset at the dielectric/barrier interface.

In addition to the energy distribution, the introduction of spatial distribution of trap states provides another time and gate voltage dependent component into the trapping/de-trapping behavior. Let us illustrate the impact of DIGS on a CV hysteresis measurement. Under forward sweep toward positive V_G, DIGS located deeper in the dielectric are progressively populated. For reverse sweep toward negative V_G, when the Fermi level moves towards the AlGaN VB, shallower DIGS near the interface are quickly emitted into the CB, however, DIGS located deeper in the dielectric remain populated due to slower tunneling-controlled emission process. The negative DIGS charge remains stored a sufficiently long time in respect to the V_G sweeping rate, which can explain unexpected CV hysteresis in the spill-over regime (discussed in more detail in Section 2.3.1).

2.2.3. Dielectric Bulk Traps and 'Border' Traps

Dielectric bulk (or more often oxide) traps (OT) are the defect states in the dielectric band gap that are able to change the charge state due to tunneling of carriers from the electrodes, i.e., semiconductor CB/VB and the gate metal. They are responsible for leakage current degradation and breakdown as well as V_{TH} instabilities in MISHEMTs and MISFETs. For oxide gate dielectrics, OTs are commonly associated with oxygen vacancies (V$_O$), being a prevalent native defect in thin oxide films deposited on semiconductors [92]. It has been predicted theoretically that V$_O$ point defect and C impurity in Al$_2$O$_3$ and HfO$_2$ show several charge states in contact with GaN and can therefore act as effective trapping states for both, electrons and holes [93,94]. Indeed, OT trapping has been suggested to affect BTI in GaN MISHEMTs [16,95] and plays a major role in BTI of MISFETs, as will be discussed in more detail in Sections 3 and 4.

In a simple approach, OT with relatively low density can be modelled as a single defect levels E_T distributed in the dielectric with activation energy given by transition between the E_T and the CB/VB. However, a more complex defect configurations can exist in amorphous dielectric layers. A detailed analysis of PBTI behavior in GaN MISFETs with SiN$_x$ and Al$_2$O$_3$ dielectrics observed by Wu et al. [27] was found to be well modeled by using the Gaussian distribution of OT in the form

$$D_{OT}(E,x) = \frac{D_{OT0}}{\sigma_t \sqrt{2\pi}} exp\left(-\frac{E-\mu_t(x)}{2\sigma_t^2}\right) \quad (5)$$

where E is the energy within the dielectric band gap, x is the spatial position inside the dielectric layer, D_{OT0} is the peak OT density, μt and σt are the mean and the standard deviations of the Gaussian distributions. The dependence of μt on x, having a form of exponential decay $exp(-x/x_0)$, translates the tunneling into the energy distribution effect. Such OT band models are exemplified in Figure 3a for SiN/GaN and Al$_2$O$_3$/GaN MISHFETs together with corresponding OT energy distributions [27]. Note that apart from mean energy position μt, also spread of the distribution σt affect the resulting PBTI, determining the accessibility of electrons from the GaN conduction band for trapping/de-trapping process at positive V_G. This empirical model can effectively describe the observed power-law dependence of V_{TH} transients on voltage and temperature during both stress and recovery periods upon PBTI [27]. Furthermore, OT band model has been shown to be effective also in description of PBTI and NBTI in GaN MISHFETs [28]. NBTI for both MISHEMTs and MISHFETs can be also affected by injection of electrons from the metal electrode into OT at the metal/dielectric interface, leading to positive V_{TH} shift under negative bias stress [16].

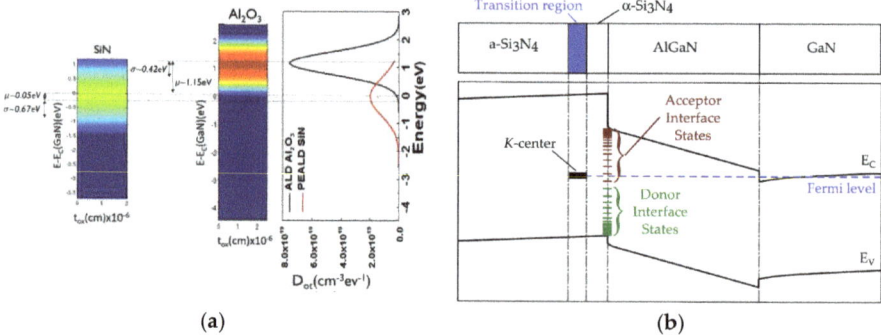

Figure 3. (a) Oxide trap (OT) defect band models in the SiN$_x$/Al$_2$O$_3$ gate dielectric (left) and the energy distribution of the gate dielectric defects for to two different gate dielectrics (right). Copyright © 2020 IEEE. Reprinted, with permission, from Ref. [27]. (b) BT model for the Fermi level pinning due to K-centers inside the silicon nitride. At the Si$_3$N$_4$/(Al)GaN interface, a U-shaped distribution of IT and one discrete donor level (BT). Copyright © 2020 AIP. Reprinted from [34], with permission from AIP.

Border traps (BTs, also called near-interface OT or slow traps) represent a special case of oxide traps, physically separated from the interface with the semiconductor. BTs, therefore, exchange the charge only with the semiconductor and with a much slower rate than IT. The physical separation of BTs can be provided by an ultra-thin (comparable to tunneling distance of the electrons) high-quality dielectric in contact with the semiconductor. The presence of a BT has been suggested by Bakeroot et al. [34], who studied 2DEG origin in GaN MISHEMTs with in-situ grown SiN dielectric. BT were proposed to form at the transition between a thin nanocrystalline and amorphous SiN layers depicted in the band diagram shown in Figure 3b. These BTs are assumed to originate from Si dangling bonds also called K-centers [96], energetically located in the middle of the band gap. Trapping/de-trapping processes associated with BT are dominantly affected by tunneling effects with weak temperature dependence and thus show some distinct features as compared to those related to IT and OT. Special distinction between BT and IT trapping/de-trapping can be observed from the measurement of CV hysteresis. While IT lead to typical stretch-out of the CV characteristics when Fermi level is moving close to semiconductor conduction band, signature of BT effect is manifested by increased hysteresis without apparent CV stretch-out. An example of such behavior has been identified by Zhu et al. [97] in Al_2O_3/AlGaN/GaN MISHEMTs.

2.2.4. Effects of Trapping Dynamics

The models discussed above can describe IT and OT occupancy at a given time instant and statistically consider Coulombic potential of the trapping states described by the value of capture cross-section. However, they do not account for effects originating from kinetics of the capture and emission processes, recently reviewed by Ostermaier et al. [23]. These effects can be explained in terms of non-radiative multi-phonon (NMP) relaxation model introduced by Henry and Lang [98] as well as cascade mechanism proposed by Lax [99] developed for IT at the SiO_2/Si interface. Change of the defect charge state leads to change of its atomic arrangement, referred to as lattice relaxation. According to the NMP model, an electron in the conduction band needs to overcome a potential barrier E_b (by acquiring vibrational energy from the lattice) in order to be captured by an empty defect. E_b is defined as the energy barrier between total electronic energy for the initial state before capture and the final state after capture. This means that capture process is also temperature activated, leading to temperature-dependent $\sigma = \sigma_0 \exp(-E_b/k_B T)$. For emission process, the electron needs to overcome the energy barrier given by $\Delta E_T + E_b$. The validity of the lattice relaxation model for dielectric/III-N interfaces has been supported by analyzing the capture cross-section of IT in MISHEMT structures with different dielectrics and barrier compositions reported by Matys et al. [18]. The observed increase in σ (ranging from 10^{-19} to 10^{-16} cm^{-2}) with the barrier lattice mismatch to GaN channel has been explained by NMP as well as cascade model. Such behavior was attributed to stronger lattice distortion at the dielectric/barrier interface due to larger strain in the barrier [18].

Multi-charge state of defects also affects the capture-emission dynamics [23]. As an example, let us assume a defect with two stable configurations after capturing one or two electrons, corresponding to states 1 and 2, respectively. Following the NMP model, in transition of the defect from state 2 to state 1 (representing electron emission), an electron needs to overcome the barrier given by the crossing point of the total electronic energy for state 2 and state 1, ΔE_{2-1}. For transfer of defect from state 1 to unoccupied state (also representing electron emission), an electron needs to overcome the barrier of $\Delta E_T + E_b$ discussed in the previous paragraph. The electron emission process from the same defect then provides two activation energies. Furthermore, the applied electric field in the oxide alter the transition barrier for both processes, leading to voltage dependent response of the emission processes. These effects are referred to as second order dynamics effect [23] and has been intensively studied for the SiO_2/Si interface [100,101]. Presence of multi-charge state IT in GaN MISHFET structure has been recently suggested by Taoka et al. [102]. The authors analyzed capture cross-section of IT in Al_2O_3/GaN MOS structures and observed much lower values of σ (in the range of 10^{-17}–10^{-19} cm^2) compared to that for SiO_2/Si interface (10^{-16}–10^{-15} cm^2), which was attributed to existence of multi-charge state nature of IT at the Al_2O_3/GaN interface [102].

2.2.5. Pictorial View of V_{TH} Instabilities in GaN MISHEMTs

Before we review recent BTI studies of GaN MISHEMTs, it is useful to outline the processes responsible for V_{TH} instabilities in these devices. Figure 4a–c depict the band diagram across the gate structure in the equilibrium (a) and under application of positive and negative V_G (b–c). In this simplified picture, only IT and OT are considered. We note that despite distinct differences, trapping/de-trapping mechanisms of OT, dielectric DIGS and BT share a common feature of tunneling effects affecting the traps occupancy. Further, trapping states in the III-N epitaxial layer are neglected. As depicted in the band diagram shown in Figure 4a, all traps below the Fermi level are occupied in the equilibrium. When certain positive bias is applied to the gate (referred in the following to as $V_{G,spill}$) in PBTI test, electrons from the channel "spill-over" into the barrier conduction band (Figure 4b). These free electrons are captured by empty IT and also by OT, when OT energy level or energy band become aligned with the barrier CB edge (Figure 4b). As a result, V_{TH} shifts into positive direction regardless of the traps' nature (i.e., donor- or acceptor-like traps). The stress-induced V_{TH} drift can be monitored electrical techniques described in the next section, giving the information on the electron capture by IT and OT. After stressing period, the recovery of the device biased at $V_{GS} = 0$ typically shows negative V_{TH} shift eventually approaching its pre-stressing value. The V_{TH} recovery transients acquired for different positive stressing bias and temperatures can be then used to characterize the trap emission kinetics [20].

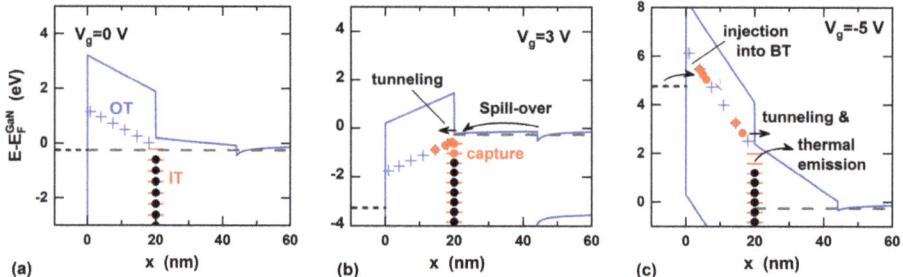

Figure 4. (a) Schematic cross-sectional band diagrams of MISHEMT under the gate showing an example of dielectric/barrier interface traps (ITs) and oxide traps (OTs) distributed in the dielectric layer in thermal equilibrium (a) and under positive (b) and negative (c) V_G applied and possible capture and emission processes inducing different V_{TH} shifts (b,c).

In typical NBTI test, negative V_{GS} is applied on the gate and stress-induces V_{TH} shift is monitored. Band diagram under negative V_{GS} is depicted in Figure 4c. Most commonly, NBTI results in negative drift of V_{TH}, which is attributed to field-enhanced electron emission from both, IT and OT located close to the dielectric/barrier interface. Similar to recovery process after PBTI, stress-induced V_{TH} shift at different stressing voltages and temperatures is used to characterize NBTI kinetics. In addition, if the OT level or band coincide with the metal Fermi level, electrons from the metal can be injected into OT, giving rise to a positive V_{TH} drift (Figure 4c). Metal electrons injection counteracts the V_{TH} drift induced by electron emission from IT and OT localized close to the interface [16]. Due to electrostatics, this effect has much weaker impact on the V_{TH} shift as compared to IT and OT capture/emission, yet, the coaction of processes depicted in Figure 4c has been reported to result in negligible apparent GaN MIS capacitor CV hysteresis at elevated temperatures [16].

Due to extremally wide range of trap time constants in GaN devices, it is practically difficult to achieve trap occupation corresponding to thermal equilibrium (such as that depicted in Figure 4a) for repetitive BTI testing. For NBTI in particular, pre-stress trap occupation has a strong impact on the resulting V_{TH} instability, as the V_{TH} drift measured on virgin sample can largely differ from that of the successive measurement. It is therefore important to establish a reference condition before any BTI investigation. Commonly used approach is the application of some de-trapping step either by application of low negative V_G stress [103] or light exposure (microscope of UV light) of the sample [28] followed by resting the sample in unbiased condition for sufficient time. The sample bake-out at 100–150 °C for 30 min is also very effective pre-testing procedure [19].

2.3. BTI Measurement Techniques

BTI evaluation can be performed using different testing methods. The simplest PBTI test represents the measurement of double-sweep I_D-V_{GS} characteristics of transistor or CV traces of large-area MIS capacitor with increasing positive gate bias, where V_{TH} hysteresis between forward and reverse sweep is evaluated. While the former characteristics are commonly performed for initial screening of the V_{TH} instability, CV measurements have been also used for deeper study of traps distribution in MISHEMT gate stacks [41,89,97]. For a more comprehensive investigation of BTI behavior in GaN transistors, a variety of stress-measure techniques has been employed. Here, positive/negative V_{GS} is applied during stressing period, which is typically followed by recovery period ($V_{GS} = 0$). During both periods, device parameters such as V_{TH}, g_m, and R_{ON} are monitored via short sampling or measurement sequences interrupting the stressing/recovery [20,28,75]. These techniques provide a comprehensive information on capture and emission kinetics of the device. Standard tests such as high temperature gate bias (HTGB) and high temperature reverse bias (HTRB), stressing the device at negative V_G and in the OFF state, have been also employed for NBTI investigation of GaN MISHEMTs [104,105]. In addition, novel NBTI test conditions tailored for GaN switching devices were developed, including High Temperature Source Current (HTSC) with semi-ON state applied during stressing period [105] and ON/OFF switching stressing [106]. Due to their importance in deeper analysis of BTI mechanisms, in the following, we will discuss some aspects of the CV hysteresis measurements of GaN MISHEMT structures and stress-measure techniques used for BTI investigation in the literature.

2.3.1. Capacitance Techniques

Despite its simple implementation, the main advantage of the capacitance techniques is that they offer straightforward information on the dynamic of charge distribution in the gate structure. When applied to large-area test diode structure, they are insensitive to parasitic surface and thermal effects compared to transistor IV measurements, reducing the data interpretation into essentially 1D problem. On the other hand, capacitance meters generally offer slower reaction time compared to current measurements (unless dedicated fast meters employed, e.g., for DLTS are used) and accuracy of capacitance measurement is limited by the onset of excessive gate leakage current. In the simplest approach, PBTI can be evaluated from double-sweep CV measurement exemplified in Figure 5 for Al_2O_3/AlGaN/GaN MIS heterostructure [107]. Typically, V_G is swept from negative to positive voltages in forward direction and backward in the reverse direction for given temperature. From set of the CV sweeps performed for increased maximum positive V_G ($V_{G,max}$), CV hysteresis at 2DEG depletion part (ΔV_{TH}) is extracted. For high-quality gate oxides, also CV hysteresis ($\Delta V_{G,spill}$) and stretch-out in the spill-over region can be evaluated [107]. Note the higher $\Delta V_{G,spill}$ compared to ΔV_{TH}. This is caused by partial emission of IT and/or OT populated at positive V_G − s during the subsequent backward sweep. Traps with shorter τ_E than the time it takes to sweep the V_G from $V_{G,max}$ to V_{TH} therefore do not contribute to apparent ΔV_{TH} [41]. In addition to a strong dependence on $V_{G,max}$, resulting ΔV_{TH} also depends on the V_G sweep rate [89]. If present, $\Delta V_{G,spill}$ is also strong function of measurement signal frequency [15,108,109] and temperature [17].

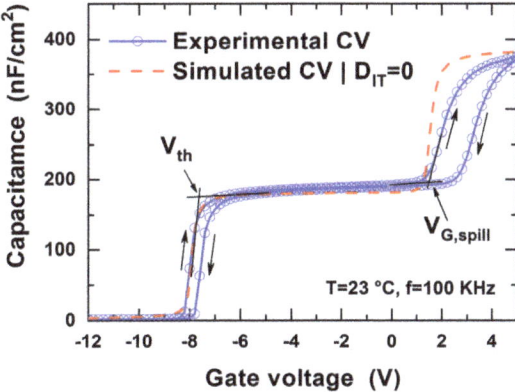

Figure 5. Example of double-sweep CV characteristic measured on large-area Al_2O_3/AlGaN/GaN MISHEMT structure together with the "ideal" CV curve, calculated using Poisson solver.

Theoretical and experimental study of D-mode Al_2O_3/AlGaN/GaN MISH structures using CV hysteresis measurements was performed by Matys et al. [89]. The authors observed gradual increase in both ΔV_{th} and $\Delta V_{G,spill}$ with $V_{G,max}$ increasing that further increased with lowering of the sweep rate. In addition, lower sweep rate resulted in stronger stretch-out of the CV curve in the spill-over regime. However, shallower IT available for capture/emission in D-mode MISHEMT are expected to result in negligible $\Delta V_{G,spill}$ as well as ΔV_{TH} with $V_{G,max}$ increasing [107]. Instead, the authors proposed the presence of DIGS in the Al_2O_3 dielectric described in Section 2.2.2 (Equations (3) and (4)). Assuming $N_0 = 8 \times 10^{17}$ eV^{-1}cm^{-3}, $x_l = 4$ nm, and $\sigma_0 = 10^{-15}$ cm^{-2}, calculated results were able to reproduce complex PBTI as well as NBTI behavior observed experimentally [89]. Somewhat different CV hysteresis behaviors of D-mode Al_2O_3/AlGaN/GaN MISH structures subjected to PBTI were observed by Zhu et al. [97]. Here, increasing of $V_{G,max}$ resulted in gradual increase of ΔV_{th} and $\Delta V_{G,spill}$, which was also accompanied with positive shift of V_{TH}, i.e., rigid right-shift of the CV curve along the V_G axis. Such behavior was attributed to dominant BT capture/emission process (described in Section 2.2.3) enhanced by IT trapping. Despite some controversy, these results suggest that traps located in the oxide close to the dielectric/barrier interface play a vital role for PBTI behavior of GaN MISHEMTs. It also demonstrates that simple CV hysteresis measurement can provide detail information on the BTI processes, complementing the IV measurements of transistors.

In the CV hysteresis measurement, availability of empty traps (located above the Fermi level in the equilibrium) plays a critical role for resulting ΔV_{th}. As discussed in Section 2.1, band bending at the dielectric/barrier interface $\Phi_{interface}$ is given by the amount of P_S compensation. For the practical comparison between the MISHEMTs, it is more appropriate to define the net interface charge $Q_{interface} = P_S + Q_{comp} + qN_{IT}$ (N_{IT} is the effective interface trapped charge), which can be extracted from the slope of $V_{TH} = f(t_{dielectric})$ [16,39]. The effect of $Q_{interface}$ on the CV hysteresis was analyzed for Al_2O_3/AlGaN/GaN large-area MISHEMT capacitors with nominally same heterostructures but different $Q_{interface}/q$ of -1×10^{13} and 1×10^{12} cm^{-2} resulting from different oxide PDA [41]. Despite relatively high D_{it} (~10^{12} eV^{-1}cm^{-2}eV) determined for both structures, devices with PDA ($Q_{interface}/q = 10^{12}$ cm^{-2}) showed negligible CV hysteresis with increasing $V_{G,max}$ (Figure 6b), while significant and enhanced CV hysteresis with increased $V_{G,max}$ was observed for structures without PDA ($Q_{int}/q = -1 \times 10^{13}$ cm^{-2}). Negligible ΔV_{TH} for structures with PDA can be explained by fast emission of shallower IT populated at forward V_G bias (see corresponding band diagram in Figure 6a) during the backward sweep, resulting in similar IT population at $V_G \sim V_{th}$ for forward and backward sweep. In contrast, electrons are captured by much deeper empty IT available in the structure without PDA (see Figure 6a) at positive V_G. This is then followed by slower IT emission during the backward sweep, thus higher negative charge stored in IT after backward measurement compared to the equilibrium. This behavior illustrates that MISHEMTs with similar interface quality but different $Q_{interface}$ can show different CV hysteresis.

Finally, it is also possible to monitor NBTI in D-mode MISHEMT structures (or PBTI in E-mode MISHEMT structures) using capacitance transient (C-t) measurements [54,110]. In this technique, transient V_{TH} drift is deduced from the C-t measurement performed at V_G bias corresponding to depletion part of the CV curve (e.g., at $V_G = V_{TH} - 0.3$ V), giving the highest sensitivity for V_{TH} shift. After optional pre-filling of the traps at positive V_G, the measured C transient is recalculated to V_{TH} transient using corresponding part of the CV curve [54,110]. The disadvantage of this technique for BTI investigation is the fundamental limitation of the stressing voltage selection and above mentioned relatively low reaction time of the capacitance measurement.

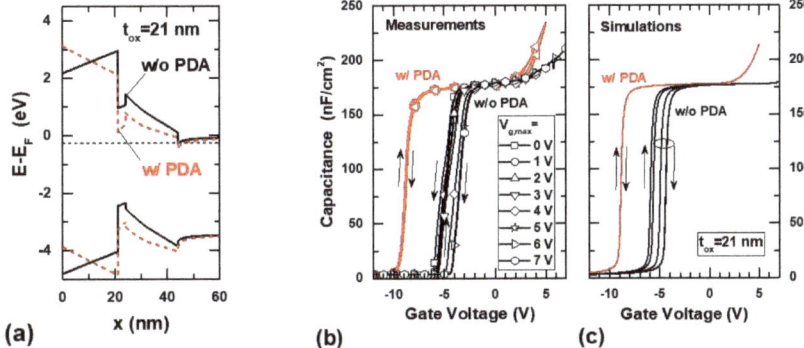

Figure 6. (**a**). Band diagrams of MISHEMT structures calculated using Poisson equation and assuming $Q_{interface}/q$ of -1×10^{13} (w/o PDA) and 1×10^{12} cm^{-2} (w/PDA). Experimental (**b**) and simulated (**c**) CV hysteresis of MISHEMT structures without and with PDA measured using different maximum V_G ranging from 0 to 7 V. Reprinted from [41], with permission from Elsevier. Copyright © 2020 Elsevier.

2.3.2. Stress-Measure IV and Pulsed I_D Techniques

Stress-measure techniques are based on monitoring of the device parameters during stressing and recovery period. The monitoring is performed by repeated interruptions of the stressing/recovery bias, during which fast DC I_D-V_{GS}, I_D-V_{DS} [21,28,104,105] or pulsed I_D [22,33,111] are measured. While the DC measurements offer detail monitoring of the change in device parameters in time (V_{TH}, R_{ON}, and g_m), pulsed I_D measurement provide only information of V_{TH} drift. On the other hand, short response time of the pulsed measurement, referred to as measure-stress-measure (MSM) technique in the literature [23], allows one to monitor capture and emission processes with sub-µs resolution, which is of particular importance in the study of trapping dynamics. Yet, some change of the trap's occupation induced during the measure period cannot be avoided and needs to be carefully considered in the design of the BTI experiments [23]. Since MSM technique has been recently employed for BTI study in MISHEMTs by several groups [22,24,25], it will be described in more detail in the next paragraph. Many researchers have also employed double-channel pulsed measurement of I_D-V_{GS} characteristics for PBTI assessment [24,75]. Here, quiescent gate bias represents the stressing V_G magnitude while the I_D-V_{GS} characteristics measured at low V_{DS} provide information on the V_{TH} drift and possible g_m change.

The MSM technique monitors V_{TH} drift using an oscilloscope base setup depicted in Figure 7a. In the measurement sequence (Figure 7b), device under test is first stressed at forward gate bias $V_{G,stress}$ for time period t_{stress} (V_{DS} = 0 V). Then, V_G is stepped to zero (V_{DS} = 0 V) and the device recovery is monitored by measurement of V_D in short moments at certain time intervals (t_{rec}) and small $V_{D,meas}$. V_{TH} drift (ΔV_{TH}) is evaluated from the transient response of V_D in respect to the virgin transfer characteristic. ΔV_{TH} can be measured for constant $V_{G,stress}$ and logarithmically increased t_{stress} or using increased $V_{G,stress}$ and constant t_{stress}. This technique allows evaluation of V_{TH} drift transients for a wide range of stressing and recovery times, as exemplified in Figure 7c,d, where V_{TH} recovery drifts for t_{stress} of 100 ns and 100 s as a function of $V_{G,stress}$ are shown. It is best suited for D-mode MISHEMTs, where stressing and measurement takes place at positive and negative V_G ($V_{TH} < V_{G,meas} < 0$), respectively. MSM technique has been also implemented by using sampling-mode DC measurements performed by, e.g., standard Keithley 4200 system [20,24,25]. Although such instruments provide much longer response time (~10 ms) compared to pulsed measurements (~100 ns), versatility of these measurement systems extends implementation of MSM technique for E-mode devices [25], application of advanced stressing conditions [105], and detailed device characterization during and immediately after the stressing [20].

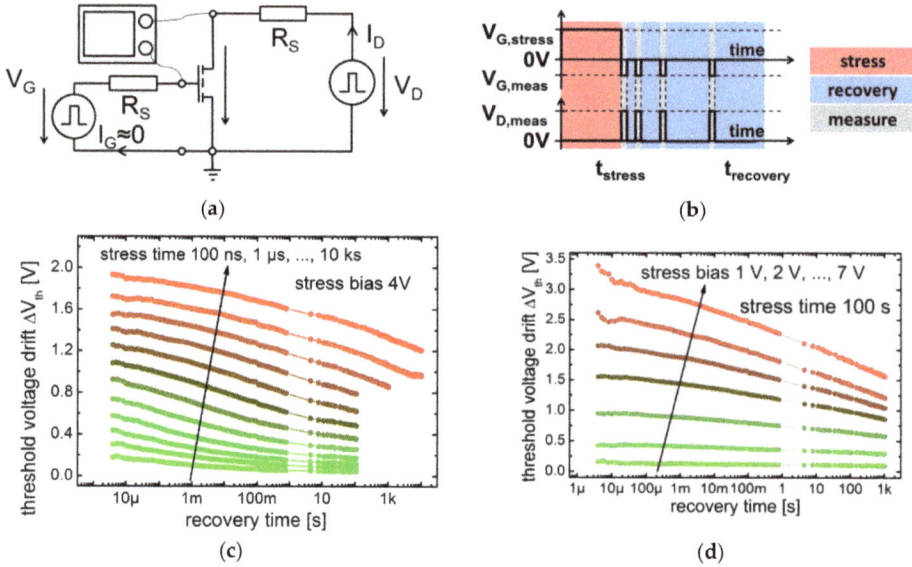

Figure 7. Measurement setup (**a**) and pulse pattern (**b**) used for stress-recovery cycle of measure-stress-measure (MSM) technique. To measure the transient response of V_D, the biases are pulsed to $V_G = V_{G,meas}$ and $V_D = V_{D,meas}$. (**c**,**d**) Example of recovery transients measured for constant $V_{G,stress} = 4$ V and varying stress times ranging from 100 ns to 100 ks (**c**) and for constant t_{stress} of 100 s while varying $V_{G,stress}$ from 1 to 7 V (**d**). Reprinted from Ref. [22], with permission from IEEE. Copyright © 2020 IEEE.

3. BTI in GaN MISHEMTs

Majority of BTI studies in GaN MISHEMTs are focused on PBTI of D-mode devices. This is because of great advancements achieved in the technology of non-recessed or partially recessed AlGaN/GaN MISHEMT switching devices. These studies allowed for deeper analysis of the V_{TH} drift mechanisms under positive bias stress, even though such devices are not expected to operate at such conditions. In fact, only a limited number of studies investigating PBTI in E-mode GaN MISHEMTs are available in the literature, which is simply because only a few design concepts of non-recessed E-mode MISHEMTs are available [13,14,43,45,46]. On the other hand, NBTI represents a major concern in the D-mode GaN MISHEMTs used, e.g., in the cascode configuration. However, due to availability of reliable and stable SB HEMTs, less interest has been given to NBTI investigations in D-mode GaN MISHEMTs. We will therefore review the most relevant research works aiming for a deeper understanding of the PBTI mechanisms in D- and E-mode GaN MISHEMTs and NBTI in D-mode MISHMETs.

3.1. PBTI in D-Mode GaN MISHEMTs

Among the first, Lagger et al. [20] studied PBTI in D-mode Al$_2$O$_3$/AlGaN/GaN MISHEMTs using MSM technique for monitoring of V_{TH} drifts upon positive bias stress and recovery. In their later research [22], extended MSM technique (improved by using an oscilloscope-based measurement) was employed to study PBTI in D-mode SiO$_2$/AlGaN/GaN MISHEMTs. Measured ΔV_{TH} drift in an extremely broad range of recovery times is exemplified in Figure 7c,d for different t_{stress} and $V_{G,stress}$. For the data interpretation, the authors considered any defects in the active energy region of the gate stack capable to exchange the charge with 2DEG, without explicit discrimination between IT and OT. The measured ΔV_{TH} is then related to change of "interface" traps occupation (ΔN_{IT}) as $\Delta N_{IT} = -C_{OX}(\Delta V_{TH}/q)$. Instead of calculating D_{IT} distribution or density of OT, ΔN_{IT} is interpreted using concept of capture emission time (CET) map, as originally proposed for analysis of NBTI in Si MOSFETs [112]. Figure 8a shows a CET map calculated from the recovery data shown in Figure 7c. Here, all trap states are described by their ΔV_{TH} (ΔN_{IT}) per decade in 3-dimensional space of corresponding capture and emission time constants, for given $V_{G,stress}$ and temperature [22]. In the case of capture/emission of an electron from/to semiconductor CB with corresponding lattice relaxation, the CET map should comprise only positive entries. However, note that CET map shown in Figure 8a includes also negative values that correspond to decrease in ΔN_{IT} with t_{stress} increasing, even though it first increases with t_{sress} from the beginning of the recovery process. The decrease in N_{IT} with increasing of t_{stress} was attributed to second order dynamics effects discussed in Section 2.2.4.

Due to existing barrier between 2DEG and IT, trap dynamic is also affected by electron transport trough the barrier layer itself. This was pointed out by Ostermaier et al. [113], who investigated V_{TH} drift of MISHEMTs with SiN gate dielectric subjected to positive bias stress at different temperatures. It was observed that the onset of positive V_{TH} drift starts at longer t_{stress} as the $V_{G,stress}$ decreases. From the thermal activation of this V_{TH} onset, E_A was found to increase linearly with $V_{G,stress}$ decreasing down to $V_{G,stress}$ = 1 V. Beyond this bias, E_A remained constant at a value of ~0.52 eV [113]. Such behavior was explained by IT capture via trap-assisted tunneling of electrons from the barrier across the triangular CB edge near the interface. This suggests that effective IT capture represents a serial process, characterized by the sum of actual defect τ_C and the time constant related to the transport process through the barrier.

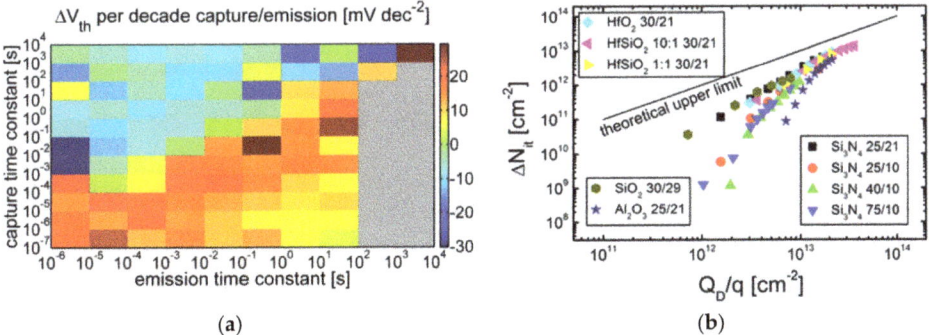

Figure 8. (a) Capture emission time (CET) map extracted from the recovery transients exemplified in Figure 7 (c,d). © 2020 IEEE. Reprinted, with permission, from Ref. [22]. (b) Measured dependence of ΔN_{IT} on the dielectric charge Q_D/q. The deviation from linear dependence between the different samples originates from the variation in $V_{G,spill}$. Reprinted from Ref. [114], with a permission of AIP Publishing. Copyright © 2020 AIP.

MSM technique in conjunction with CET map data analysis was used to study MISHEMTs with various dielectric materials [114]. Although different gate materials resulted in different ΔV_{TH} for given t_{stress}, the authors pointed out that also gate electrostatics, affecting availability of free electrons for capture, need to be considered in such comparison. This is demonstrated in Figure 8b showing the dependence of ΔN_{IT} on the gate displacement charge ($Q_D = C_D \times V_G$, C_D is the dielectric capacitance), which converges towards the same dependence for all dielectric materials, reaching its upper limit $\Delta N_{IT} = Q_D$ [114]. Note that higher the C_D, the more charges are accumulated at the interface for given positive V_G. The linear dependence of ΔN_{it} on Q_D/q without apparent saturation observed for $V_G > V_{G,spill}$ in all dielectrics studied in [114] indicates that a higher density of traps (IT and OT) is available than the number of free electrons present in the reservoir, i.e., 2DEG in bare AlGaN/GaN heterojunction. A similar situation was concluded also in the study of Winzer et al. [111], investigating PBTI in D-mode MISHEMTs with Al_2O_3 and HfO_2 gate dielectrics with similar C_D. Here, ΔN_{IT} as high as ~10^{13} cm^{-2} was determined from positive bias stress at $V_G > V_{G,spill}$ for both devices. However, 2DEG concentration of 8×10^{12} cm^{-2} was determined by the Hall measurements of the AlGaN/GaN heterostructure. As a consequence, such high density of interface traps limits their evaluation by electrical methods. Further, when an MISHEMT is driven into spill-over regime, voltage drop across the barrier remains unchanged with further V_G increasing. Additional increase if V_G results in voltage drop solely across the dielectric and is limited by the dielectric critical breakdown field. Ostermaier at el. [23] applied these two limitations for definition of the practical lifetime requirements for E-mode MISHEMTs. First, sufficiently low ΔV_{TH} assuring maximum specified R_{ON} and the minimum specified I_D at the end-of-life must be fulfilled at the operating V_G. Second, the stability of the gate dielectric must be assured at maximum V_G at end-of-life. Using the experimental values of critical electric field of commonly applied dielectrics, maximum ΔN_{it} was estimated to be in the range of 4–8×10^{12} cm^{-2} for typical 2DEG sheet channel density of 0.5–1×10^{12} cm^{-2} [23].

The effect of the gate dielectric growth method on PBTI was also studied by Meneghesso et al. [75], who reported comprehensive reliability investigation of partially recessed barrier MISHEMTs with SiN gate dielectric grown by rapid thermal CVD (RTCVD) and plasma-enhanced ALD (PEALD). As deduced form the pulsed I_D-V_{GS} and MSM measurement, MISHEMTs with PEALD dielectric showed notably lower V_{TH} drift with resulting ΔN_{IT} of ~3.5×10^{11} cm^{-2} (for t_{stress} = 1000 s) compared to devices with RTCVD dielectric showing ΔN_{IT} of ~2×10^{12} cm^{-2}. Interestingly, the recovery times for both devices exceeded 1000 s for $V_{G,stress}$ = 2.5 V. V_{TH} drift was found to correlate with the gate leakage current at positive gate voltages, indicating that bulk OTs, which are likely to govern the gate leakage mechanism, are also responsible for V_{TH} drift upon positive stressing bias. In addition, the gate robustness under forward bias was examined using step-stress and time-dependent dielectric breakdown (TDDB) measurements [75]. Improved gate robustness of PEALD grown SiN gate dielectric as compared to RTCVD SiN has been attributed to lower OT density in this dielectric due to lower probability of the percolation leakage path formation. These results highlight the importance of the gate dielectric technology, as non-optimal growth conditions leading to higher density of OT can result in inferior device reliability in terms of V_{TH} stability as well as gate robustness.

Due to presence of different defects in the gate stack and variability of their capture/emission dynamics affected also by the transport through the barrier discussed above, it is generally challenging to discriminate, which defects dominantly affects the PBTI behavior in MISHEMTs. Yet, detailed analysis of the V_{TH} drift induced by forward bias stress and/or subsequent recovery can provide indications on the nature of the relevant defects. Among others, Zhang et al. [24] investigated PBTI in MISHEMTs with PECVD grown SiN gate dielectric using step stress-recovery experiment. The capture process was found to follow a two-step trapping process with fast electron trapping (with time constant below 100 ms) followed by a slow dynamic, featuring a logarithmic time-dependent V_{TH} drift. The fast capture process was attributed to IT capture, while the slow capture process was proposed to result from population of OT located close to the interface. Alternatively, the authors considered the importance of the electrostatic feedback effect, where electron capture rises the barrier potential leading to the decrease of the gate current and thus availability of free electrons. In the study of Wu et al. [103], V_{TH} recovery after forward bias stressing at different temperatures was monitored via fast DC I_D-V_{GS} measurement in AlGaN/GaN MISHEMTs with ALD Al$_2$O$_3$/in-situ SiN gate dielectric. The derivative of V_{TH} transients revealed distinct peaks, which indicates the presence of discrete traps level. The trap level E_A of 0.69–0.7 eV was determined and ascribed to donor-like Si$_3$N$_4$/AlGaN IT level. Although the presented results seem to be controversial, they highlight the importance of the gate dielectric technology on the defect nature and density. While the MOCVD grown in-situ SiN is expected to provide high-quality interface with the AlGaN barrier, SiN deposited by PECVD may be expected to result in the gate stack with relatively higher IT density as well as bulk OT.

3.2. PBTI in E-Mode MISHEMTs

As discussed in Section 2.1, E-mode operation of GaN MISHEMTs can be achieved, e.g., by the barrier fluorination [14] or polarization engineering approach [13]. Investigation of PBTI in E-mode MISHEMTs with fluorination approach was performed by Wu et al. [115]. Under gate bias stress performed at different positive V_{GS} (4–6 V) and temperatures (30–150 °C) for t_{stress} of 10 ks, the authors observed positive V_{TH} shift up to 1 V with time evolution, which can be fitted by the empirical power-law dependence [27]

$$\Delta V_{TH} = A_0 (V_{GS} - V_{TH})^\gamma t_{stress}^n \quad (6)$$

where A is the pre-factor and n is the time exponent. This behavior will be further discussed in more detail in Section 4.1. Similar results were observed also for devices without fluorination, so that the PBTI was attributed to pre-existing IT and OT, while negligible plasma treatment-induced trap generation was concluded.

Up to now, we have only discussed PBTI behavior with positive/negative V_{TH} drift under stressing/recovery. However, also opposite V_{TH} drift has been reported for E-mode devices with polarization-engineered barrier structure. In our recent study [25], PBTI in Al$_2$O$_3$/InGaN/AlGaN/GaN MISHEMTs was investigated using the MSM technique. In this structure, a high negative polarization charge at the InGaN/AlGaN interface raises the bands, leading to formation of 2-dimensional hole gas (2DHG) at this interface [13]. Typical MSM measurements are shown in Figure 9a for stress (left, t_{stress} = 100 ms–50 s) and recovery (right, t_{rec} = 50 s) period. The devices show negative V_{TH} drift during stress period and positive V_{TH} drift during recovery. Yet, the final V_{TH} transient ends-up at higher value in respect to the beginning of the stressing. Such behavior was explained by the following model (Figure 9b): When positive $V_{G,stress}$ is applied to the gate, electrons accumulated at AlGaN/GaN interface are injected into the InGaN layer and recombine with holes in 2DHG, emitting photons with energy of around 3 eV (i.e., the bandgap of the InGaN). Some of these photons are absorbed by the Al$_2$O$_3$/InGaN interface states and the released electrons can then tunnel through the triangular barrier of the oxide CB edge. The emptied states build up positive charge, which results in negative V_{TH} shift. During recovery, structure tends to return to the equilibrium, however, some holes are depleted from the 2DHG. Therefore, resulting V_{TH} is shifted to more positive values as compared to pre-stress condition.

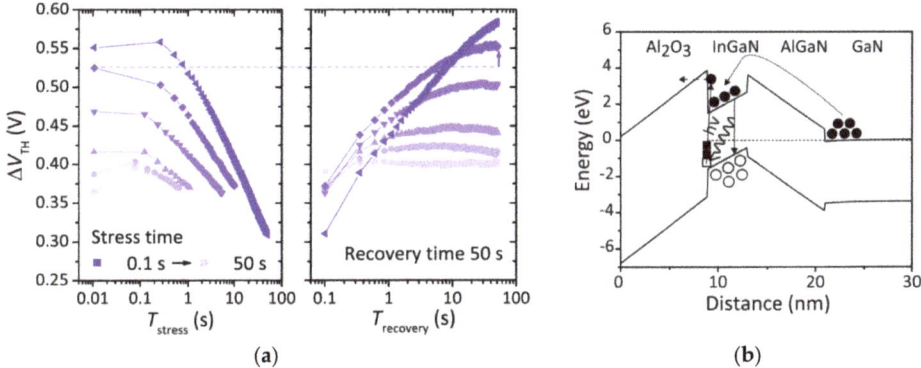

Figure 9. (a) Successive MSM measurement of Ni/Al$_2$O$_3$/InGaN/AlGaN/GaN MOS HEMT (stress—left, recovery—right). Measurements start with the shortest stressing time (0.1 s) and continue with increased stressing time. (b) Schematic representation of selected processes during stress. Electrons are injected from 2DEG into InGaN. Electrons recombine with holes in InGaN and emit photons that, in turn, de-trap interface states. Emitted electrons from interface states tunnel into conduction band (CB) through triangular barrier. Copyright © 2020 Elsevier. Reprinted, with permission, from Ref. [25].

3.3. NBTI in GaN MISHEMTs

Investigation of NBTI in D-mode MISHEMTs is sparse in the literature. Dalcanale et al. [105] presented an interesting NBTI investigation of GaN MISHEMTs designed to work in a cascode configuration [116]. In addition to the application of high temperature reverse bias (HTRB) testing protocol with $V_{GS} < V_{TH}$ and V_{DS} biased up to 750 V, the authors defined a novel high temperature source current (HTSC) stressing condition [105]. In this test, the device is stressed in the semi-on state operation so that hot-electrons are generated in the transistor's channel. It was found that no significant V_{TH} drift was observed under HTSC conditions, while HTRB stress results in a strong negative V_{TH} drift. This behavior was ascribed to steady-state population of IT as well as GaN buffer traps by the hot-electrons under HTSC at the drain side of the gate. On the other hand, dominant emission from these traps took place in the case of HTRB with higher V_{GS}.

4. BTI in GaN MISFETs

An often-used approach to process lateral E-mode GaN switching transistor is to fully recess the AlGaN barrier layer under the gate, while existing 2DEG in the source-to-gate and gate-to-drain regions provide a low access resistance. The MIS gate structure of fully recessed GaN devices resembles that of Si MOSFETs and many approaches developed for BTI investigation in Si devices have been adopted also for description of PBTI and NBTI issues of GaN MISFETs. Universal recovery model developed for Si MOSFETs [117] have been widely employed to describe the PBTI [27] as well as NBTI [28] behavior of GaN MISFETs. Although dielectric/GaN IT plays some role, there seems to be a general agreement that dielectric OTs have dominant impact on the PBTI and NBTI mechanism.

4.1. PBTI in GaN MISFETs

Comprehensive analysis of PBTI in fully recessed GaN MISFETs with PEALD SiN and ALD Al$_2$O$_3$ gate dielectric was performed by Wu et al. [27]. Due to observed lack of correlation between I_D-V_{GS} hysteresis (thus ΔV_{TH}) and the D_{IT} distribution measured by G-ω method, the authors employed PBTI stress-recovery tests using the MSM technique. It was found that ΔV_{TH} transients for different $V_{G,stress}$ (Figure 10a,b) can be fitted using Eqn. 6 in the whole range of t_{stress} with time exponent n in the range of 0.1–0.02. Despite notably higher D_{IT}, devices with Al$_2$O$_3$ showed about 10-times lower ΔV_{TH} compared to those with SiN, when benchmark at t_{stress} = 2 s depicted by the black arrows in Figure 10a,b. The voltage exponent γ of 1 and 2 and E_a of 0.57 and 1.02 eV was observed for devices with SiN and Al$_2$O$_3$, respectively. The recovery ΔV_{TH} transients obeyed empirical model of universal relaxation [117]

$$\Delta V_{TH}(t_{stress}, t_{relax}) = R(t_{stress}, t_{relax} = 0) r(\xi) + P(t_{stress}) \quad (7)$$

$$r(\xi) = \frac{1}{1 + B\xi^\beta} \quad (8)$$

where R and P represent recoverable and permanent degradation ascribed to different types of defects, t_{relax} is measured form the end of last stress phase, $\xi = t_{relax}/t_{stress}$ is the universal relaxation time, B is the scaling parameter and exponent β represents the dispersion parameter. From the fitting of Equations (7) and (8) to the ΔV_{TH} recovery transients (Figure 10c), devices with Al$_2$O$_3$ gate dielectric were found to show lower recoverable and permanent degradation and faster dielectric defect discharge. Such behavior was attributed to presence of OT Gaussian distributions depicted in Figure 3a. In the case of SiN, a wider distribution of OT levels ($\sigma \sim 0.67$ eV), centered below the conduction band of GaN (E_C−0.05 eV) are easily accessible by the channel carriers already at a low $V_{G,stress}$. In contrast, Al$_2$O$_3$ gate dielectric was proposed to feature a narrower distribution of OT ($\sigma \sim 0.42$ eV) located far from the conduction band edge of GaN (E_C + 1.15 eV), explaining the improved PBTI behavior in devices with ALD grown Al$_2$O$_3$ gate dielectric compared to those with SiN.

Dominant impact of dielectric OT on the V_{TH} instabilities has been proposed also by other studies. Bisi et al. [118] studied PBTI stress-recovery kinetics in Al$_2$O$_3$/GaN MIS capacitors grown by in-situ MOCVD by means of combined I_G transient, CV and capacitance MSM technique. In the low-field (oxide electric field <3.3 MV/cm) regime, I_G stress and recovery transients were found to obey power-low $I_G(t) \propto t^{-\alpha}$ with $\alpha \sim 1$, suggesting trapping and de-trapping of near-interface OT. In contrast, high-field regime (>3.3 MV/cm) was characterized by the onset of the gate leakage current promoted by OT and significant positive flat-band voltage (V_{FB}) shift, suggesting enhanced charge trapping of OT, as also revealed by very slow recovery transients [118]. Acurio et al. [119] studied PBTI in a fully recessed-gate MSIFET with PECVD SiO_2 gate dielectric using I_D-V_{GS} measurements interrupting the stressing and recovery. Similar to previous results, power-law dependence of ΔV_{TH} on t_{stress} was observed. Furthermore, trapping rate (evaluated as $\partial \log V_{TH}/\partial \log t$) exhibited a universal decreasing behavior as a function of the number of filled traps. Stress-induced ΔV_{TH} was fully recovered by applying a small negative voltage and the recovery dynamics (monitored in time window between 1 s and thousands of seconds) was found to be well described by the superimposition of two exponential functions. These emission processes were associated with two different OT. The slower trap revealed E_A of 0.93 eV and the faster trap exhibits large spread in E_A ranging from 0.45 eV to 0.82 eV. V_{TH} recovery with two different time constants have been also observed also by Iucolano et al. [120]. By measuring the hysteresis of I_D-V_{GS} characteristics with different $V_{GS,max}$ (V_{DS} = 0.1 V), partial V_{TH} recovery was reached after a few seconds, however, a complete V_{TH} recovery required more than one day of unbiased storage. Based on the numerical simulations, the fast and slow recovery processes were associated with the emission from IT and OT localized above the GaN conduction band energy, respectively.

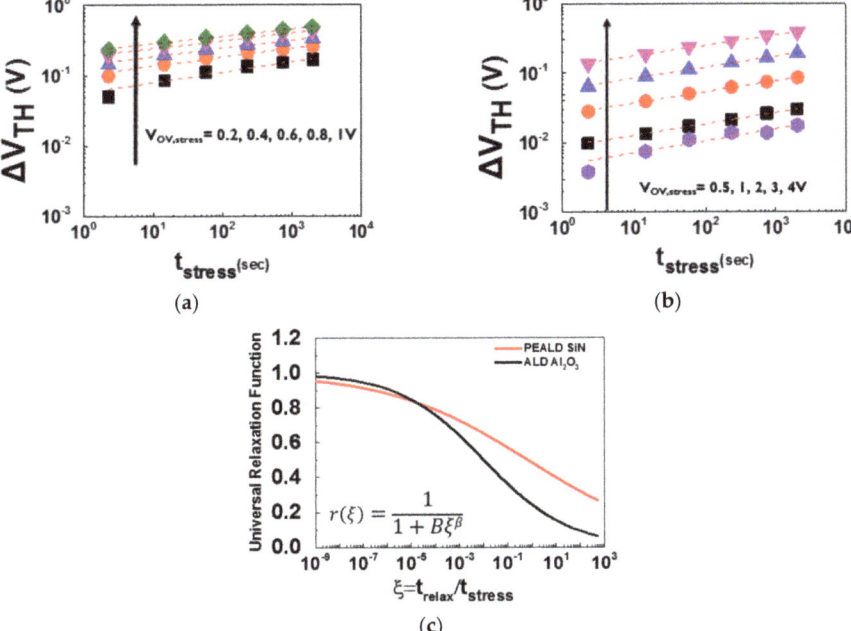

Figure 10. ΔV_{TH} transients for different t_{stress} of fully-recessed MISFETs with (**a**) plasma-enhanced ALD (PEALD) SiN and (**b**) ALD Al_2O_3 gate dielectric in a logarithmic–logarithmic scale. Dashed lines represent the power-law fits (Equation (6)) to the data. (**c**) Recovery transients fitted with the universal relaxation model (Equations (7) and (8)) showing faster relaxation for the device with Al_2O_3 gate dielectric. Copyright © 2020 IEEE. Reprinted, with permission, from Ref. [27].

In fully recess-gate MISFETs, the drain-edge of the gate terminal represent another critical region concerning the BTI, as trapping states formed close to interface between the dielectric and the barrier side-wall can negatively affect V_{TH} as well as R_{ON} stability. In the study of Chini and Iucolano [121], E-mode GaN MISFETs were subjected to PBTI in the switching-mode operation and V_{TH} and R_{ON} drift was monitored simultaneously using special designed pulsed setup [121]. Apart from positive V_{TH} drift related to IT traps under the gate region, also V_{TH} drift linked to a localized trapping in the drain-edge of the gate terminal was identified. Further, the observed increase in R_{ON} was associated with a hole-emission process taking place in the gate-drain access region within the C-doped buffer layer.

4.2. NBTI in GaN MISFETs

Current understanding of NBTI in E-mode GaN MISFETs is quite limited and some controversy exist in the observed behavior. In one of the first studies, Sang et al. [122] compared V_{TH} drifts under negative V_G stress in the D-mode MISHEMTs and E-mode GaN MISFETs with ALD Al_2O_3 gate dielectric. For the latter, the positive stress-induced V_{TH} shift was observed, which was attributed to metal gate electron injection into OT and following redistribution of the trapped charge towards the GaN channel via trap-assisted tunneling. Later, Guo and del Alamo [19] performed a more detailed investigation of NBTI in SiO_2/Al_2O_3/GaN MISFETs subjected to negative V_{GS} stress with different amplitude, duration, and temperatures. Stress-induced V_{TH} shift was found to progress through three regimes. Under low-stress (low $V_{GS,stress}$, low T, short t_{stress}), ΔV_{TH} was negative and recoverable due to electron de-trapping from pre-existing OT. Under mid-stress (low $V_{GS,stress}$, high T, longer t_{stress}), positive and recoverable ΔV_{TH} was observed, i.e., behavior similar to that reported by Sung et al. [122]. However, the cause of this effect was attributed to electron tunneling from VB to trap states in the GaN channel under the gate edges, also referred the as Zener trapping in the literature [123]. For high-stress ($V_{GS,stress} < -30$ V, RT), non-recoverable negative ΔV_{TH} was observed and ascribed to generation of new IT.

Recently, Guo and del Alamo [28] presented a comprehensive study of BTI in GaN MISFETs under moderate positive and negative gate bias stress ($V_{G,stress}$ = +5 and = 5 V) using fast I_D-V_{GS} measurements interrupting the stressing. V_{TH} evolution was monitored during the stressing and recovery phase, followed by full recovery of V_{TH} to pre-stress value. For positive $V_{G,stress}$, positive ΔV_{TH} drift was observed, which increased with stressing voltage. Nearly symmetrical behavior with negative V_{TH} drift was observed for NBTI. For both stress conditions, the V_{TH} time evolution was found to follow power-law model (Equation (6)) during the stressing and universal relaxation model (Equations (7) and (8)) during the recovery. Therefore, the authors proposed that NBTI and PBTI are caused by the same mechanism, which is the electron trapping/de-trapping in preexisting OT that form a defect band close to the dielectric/GaN interface [28]. The authors assumed the defect band extending the energies above the GaN CB edge and below the surface Fermi level at V_{GS} = 0 V. This means that some trap states are empty while some are populated with electrons at V_{GS} = 0 V. During the stress phase, the electron occupation of OT increases or decreases depending on the sign of $V_{G,stress}$, resulting in V_{TH} in positive or negative direction, respectively. In the recovery process, the trap occupation returns to the state corresponding to V_{GS} = 0 V.

Apparent positive stress-induced V_{TH} drift under NBTI was observed also in studies of Hua et al. [73,124] and He et al. [125] in E-mode GaN MISFETs with LPCVD-SiN_x/PECVD-SiN_x/GaN gate stack. The PECVD SiN_x interfacial layer (thickness of ~2 nm) grown at low temperature was employed to improve the gate stability and reliability [73]. NBTI was performed at $V_{GS,stress} = -30$ V (V_{DS} = 0 V) at temperatures of 25 and 150 °C [73,124]. While relatively low positive V_{TH} drift (<0.2 V) was observed at 25 °C, it increased to ~0.4 V for stressing at 150 °C. The positive V_{TH} drift was ascribed to metal electrode injection into OT at negative V_{GS}. In the upcoming work of this group [74], V_{TH} stability under OFF-state step-stress in similar devices was compared for different V_{GS} (0 and −20 V) applied during the stressing, using the same gate-to-drain voltage (V_{GD}). Similar to a previous study, relatively low and recoverable positive V_{TH} drift was observed for step-stress with V_{GD} up to 200 V and V_{GS} = 0 V. However, a substantially larger V_{TH} drift (~2 V) appeared for $V_{GD} > 100$ V when more negative V_{GS} was applied during the step-stress. The larger V_{TH} drift was explained by a hole-induced degradation model. Here, holes generation via impact ionization [126] or Zener trapping [123] is assumed in the high-field gate-to-drain region in the OFF-state. For stressing with negative V_{GS}, the generated holes can flow to the gate and are assumed to generate new OT in the gate dielectric, similar to TDDB mechanism [127]. The effect of holes generation on the apparent positive V_{TH} drift was further confirmed by UV light illumination of the devices subjected to OFF-state stressing [128]. However, the observed stress-induced positive shift of V_{TH} was ascribed to electron trapping during the measurement of I_D-V_{DS} characteristic, interrupting the stressing. This clearly illustrates the advantage of the MSM techniques, where V_{TH} drift is sampled quickly after the stress interruption, rather than extracted from slower measurement of the I_D-V_{GS} characteristic.

To mitigate the reverse-bias induced gate degradation in SiN$_x$/GaN MISFETs, Hua et al. [129] recently processed the transistor with channel converted from GaN to crystalline GaO$_x$N$_{1-x}$ under the gate. The oxynitride with higher bandgap (4.1 eV) compared to GaN (3.4 eV) provides also valence band offset in respect to GaN (0.6 eV), which acts as an energy barrier for holes. This barrier effectively suppresses the injection of the generated holes into the gate, improving the gate stability and robustness [128]. Robustness of V_{TH} stability upon reverse-bias stress in SiN$_x$/GaO$_x$N$_{1-x}$/GaN MISFETs can be further enhanced by varying the substrate termination [130].

The presented studies clearly point to dominant effect of OT on PBTI as well as NBTI in GaN MISFETs. This means that most commonly used dielectric materials (Al$_2$O$_3$, SiO$_2$, SiN) with research [27,28,73,118–122,124,129] as well as industry [131] graded quality contain relatively high density of OT. Apart from BTI issues, these defects represent the concern also in relation to TDDB. However, further research work focusing on the enhanced MSM technique, which allows a detailed study of capture/emission processes in the μ-s range is clearly necessary. Detailed knowledge of the OT origin can, in turn, facilitate their effective suppression via optimization of the dielectric growth technologies.

5. Conclusions and Prospects

This review illustrates that intensive research effort has been dedicated towards development of high-quality dielectric/III-N interface technology as well as deeper understanding of the BTI phenomena in GaN switching transistors. Yet, BTI currently represents a limiting reliability issue of GaN MISHEMTs and MISFETs and further technology improvements are necessary for stable operation of these devices. The observed BTI results from various trap states present in the MIS gates, including dielectric/III-N interface traps and traps distributed in the dielectric bulk (OT, BT, DIGS). In particular, PBTI represents a major concern in D-mode GaN MISHEMTs. Detail analysis of PBTI using enhanced MSM technique suggest existence of 2nd order dynamic effects as well as impact of the barrier conductance on the electron capture process. Despite a clear impact of the gate dielectric material and technology, PBTI seems to follow a common behavior when the gate electrostatic is considered, indicating that density of available traps at the interface is higher than the density of free carrier available in the 2DEG. It is therefore imperative to limit the maximum positive gate bias for D-mode MISHEMTs, so that V_{TH} drift as well as TDDB effects are mitigated. Current understanding of PBTI mechanisms in E-mode GaN MISHEMTs is limited. It seems to be clear, however, that the increased interface potential present in these devices facilitate trapping/de-trapping of very deep traps, which would be inaccessible (and thus benign) in the D-mode counterparts. In respect to PBTI requirements, this effect may limit the applicability of E-mode MISHEMTs for power switching applications.

In the case of GaN MISFETs, further study of PBTI and NBTI behavior in sub-ms time scale is necessary. On the other hand, similar behavior of PBTI has been observed by several groups, with the same empirical models used to describe the observed V_{TH} drift upon stressing and recovery period (Equations (6)–(8)). These results suggest the dominant effect of OT on PBTI as well as NBTI in GaN MISFETs. Currently, it seems that an effective way to mitigate BTI in GaN MISFETs is to optimize the growth of the dielectric, so that formation of the oxide defects is suppressed. However, there is still an open question on the origin of donor states (Q_{comp}), which compensates the GaN surface (spontaneous) polarization charge. Apparent charge neutrality of the dielectric/GaN interface (c.f. Figure 1d) therefore points to the model assuming fixed charge formed by the donor levels located between the dielectric and GaN CB edges [35,38,39]. Nevertheless, considering the current understanding of PBTI in GaN

MIS transistors, the fully recess-gate MISFETs seem to be the preferred concept for E-mode switching device over the MISHEMTs, even though the latter offer slightly lower R_{ON}.

Funding: This research was funded by the Ministry of Education, Science, Research and Sports and the Slovak Academy of Sciences, grant number VEGA 2/0109/17 and VEGA 2/0100/21.

Conflicts of Interest: The authors declare no conflict of interest. The funders had no role in the design of the study; in the collection, analyses, or interpretation of data; in the writing of the manuscript, or in the decision to publish the results.

References

1. Dmitriev, V.A.; Irvine, K.G.; Carter, C.H., Jr.; Kuznetsov, N.I.; Kalinina, E.V. Electric breakdown in GaN p–n junctions. *Appl. Phys. Lett.* **1996**, *68*, 229–231. [CrossRef]
2. Gelmont, B.; Kim, K.; Shur, M. Monte Carlo simulation of electron transport in gallium nitride. *J. Appl. Phys.* **1993**, *74*, 1818–1821. [CrossRef]
3. Sichel, E.K.; Pankove, J.I. Thermal conductivity of GaN, 25-360 K. *J. Phys. Chem. Solids* **1977**, *38*, 330. [CrossRef]
4. Mishra, U.K.; Parikh, P.; Wu, Y.-F. AlGaN/GaN HEMTs-an overview of device operation and applications. *Proc. IEEE* **2002**, *6*, 1022–1031. [CrossRef]
5. Dora, Y. High breakdown voltage achieved on AlGaN/GaN HEMTs with integrated slant field plates. *IEEE Electron Dev. Lett.* **2006**, *27*, 713–715. [CrossRef]
6. Uemoto, Y.; Hikita, M.; Ueno, H.; Matsuo, H.; Ishida, H.; Yanagihara, M.; Ueda, T.; Tanaka, T.; Ueda, D. Gate injection transistor (GIT): A normally-off AlGaN/GaN power transistor using conductivity modulation. *IEEE Trans. Electron Dev.* **2007**, *54*, 3393–3399. [CrossRef]
7. Kuzuhara, M.; Tokuda, H. Low-loss and high-voltage III-nitride transistors for power switching applications. *IEEE Trans. Electron Dev.* **2015**, *62*, 405–413. [CrossRef]
8. Rosina, M. GaN and SiC power device: Market overview. In Proceedings of the Semicon Europa, Munich, Germany, 13–16 November 2018.
9. Hashizume, T.; Nishiguchi, K.; Kaneki, S.; Kuzmik, J.; Yatabe, Z. State of the art on gate insulation and surface passivation for GaN-based power HEMTs. *Mat. Sci. Semicond. Process.* **2018**, *78*, 85–95. [CrossRef]
10. Downey, B.P.; Meyer, D.J.; Roussos, J.A.; Katzer, D.S.; Ancona, M.G.; Pan, M.; Gao, X. Effect of gate insulator thickness on RF power gain degradation of vertically scaled GaN MIS-HEMTs at 40 GHz. *IEEE Trans. Dev. Mat. Reliab.* **2015**, *15*, 474–477. [CrossRef]
11. Saito, W.; Takada, Y.; Kuraguchi, M.; Tsuda, K.; Omura, I. Recessed-gate structure approach toward normally off high-voltage AlGaN/GaN HEMT for power electronics applications. *IEEE Trans. Electron Dev.* **2006**, *53*, 356–362. [CrossRef]
12. Capriotti, M.; Fleury, C.; Bethge, O.; Rigato, M.; Lancaster, S.; Pogany, D.; Triedel, E.-B.; Hilt, O.; Brunner, F.; Würfl, J. E-mode AlGaN/GaN True-MOS, with high-k ZrO_2 gate insulator. In Proceedings of the 2015 45th European Solid-State Device Research Conference (ESSDERC), Graz, Austria, 14–18 September 2015; pp. 60–63.
13. Gregušová, D.; Blaho, M.; Haščík, Š.; Šichman, P.; Laurenčíková, A.; Seifertová, A.; Dérer, J.; Brunner, F.; Würfl, J.H.; Kuzmík, J. Polarization-engineered n^+GaN/InGaN/AlGaN/GaN normally-off MOS HEMTs. *Phys. Stat. Solidi A* **2017**, *214*, 1700407. [CrossRef]
14. Zhang, Y.; Sun, M.; Joglekar, S.J.; Fujishima, T.; Palacios, T. Threshold voltage control by gate oxide thickness in fluorinated GaN metal-oxide-semiconductor high-electron-mobility transistors. *Appl. Phys. Lett.* **2013**, *103*, 033524. [CrossRef]
15. Hori, Y.; Yatabe, Z.; Hashizume, T. Characterization of interface states in Al_2O_3/AlGaN/GaN structures for improved performance of high-electron-mobility transistors. *J. Appl. Phys.* **2013**, *114*, 244503. [CrossRef]
16. Ťapajna, M.; Jurkovič, M.; Válik, L.; Haščík, Š.; Gregušová, D.; Brunner, F.; Cho, E.-M.; Kuzmík, J. Bulk and interface trapping in the gate dielectric of GaN based metal-oxide semiconductor high-electron-mobility transistors. *Appl. Phys. Lett.* **2013**, *102*, 243509. [CrossRef]
17. Shih, H.-A.; Kudo, M.; Suzuki, T. Gate-control efficiency and interface state density evaluated from capacitance-frequency-temperature mapping for GaN-based metal-insulator-semiconductor devices. *J. Appl. Phys.* **2014**, *116*, 184507. [CrossRef]

18. Matys, M.; Stoklas, R.; Kuzmik, J.; Adamowicz, B.; Yatabe, Z.; Hashizume, T. Characterization of capture cross sections of interface states in dielectric/III-nitride heterojunction structures. *J. Appl. Phys.* **2016**, *119*, 205304. [CrossRef]
19. Guo, A.; del Alamo, J.A. Negative-bias temperature instability of GaN MOSFETs. In Proceedings of the 2016 IEEE International Reliability Physics Symposium (IRPS), Pasadena, CA, USA, 17–21 April 2016.
20. Lagger, P.; Ostermaier, C.; Pobegen, G.; Pogany, D. Towards understanding the origin of threshold voltage instability of AlGaN/GaN MIS-HEMTs. In Proceedings of the 2012 International Electron Devices Meeting, San Francisco, CA, USA, 10–13 December 2012.
21. Meneghesso, G.; Meneghini, M.; De Santi, C.; Ruzzarin, M.; Zanoni, E. Positive and negative threshold voltage instabilities in GaN-based transistors. *Microelectron. Reliab.* **2018**, *80*, 257–265. [CrossRef]
22. Lagger, P.; Reiner, M.; Pogany, D.; Ostermaier, C. Comprehensive study of the complex dynamics of forward bias-induced threshold voltage drifts in GaN based MIS-HEMTs by stress/recovery experiments. *IEEE Trans. Electron Dev.* **2014**, *61*, 1022–1030. [CrossRef]
23. Ostermaier, C.; Lagger, P.; Reiner, M.; Pogany, D. Review of bias-temperature instabilities at the III-N/dielectric interface. *Microelectron. Reliab.* **2018**, *82*, 62–83. [CrossRef]
24. Zhang, K.; Wu, M.; Lei, X.; Chen, W.; Zheng, X.; Ma, X.; Hao, Y. Observation of threshold voltage instabilities in AlGaN/GaN MIS HEMTs. *Semicond. Sci. Technol.* **2014**, *29*, 075019. [CrossRef]
25. Pohorelec, O.; Ťapajna, M.; Gregušová, D.; Gucmann, F.; Hasenöhrl, S.; Haščík, Š.; Stoklas, R.; Seifertová, A.; Pécz, B.; Tóth, L.; et al. Investigation of interfaces and threshold voltage instabilities in normally-off MOS-gated InGaN/AlGaN/GaN HEMTs. *Appl. Surf. Sci.* **2020**, *528*, 146824. [CrossRef]
26. Roccaforte, F.; Greco, G.; Fiorenza, P.; Iucolano, F. An overview of normally-off GaN-based high electron mobility transistors. *Materials* **2019**, *12*, 1599. [CrossRef] [PubMed]
27. Wu, T.-L.; Franco, J.; Marcon, D.; De Jaeger, B.; Bakeroot, B.; Stoffels, S.; Van Hove, M.; Groeseneken, G.; Decoutere, S. Toward understanding positive bias temperature instability in fully recessed-gate GaN MISFETs. *IEEE Trans. Electron Dev.* **2016**, *63*, 1853–1860. [CrossRef]
28. Guo, A.; del Alamo, J.A. Unified mechanism for positive- and negative-bias temperature instability in GaN MOSFETs. *IEEE Trans. Electron Dev.* **2017**, *64*, 2142–2147. [CrossRef]
29. del Alamo, J.A.; Lee, E.S. Stability and Reliability of Lateral GaN Power Field-Effect Transistors. *IEEE Trans. Electron Dev.* **2019**, *66*, 4578–4590. [CrossRef]
30. Ambacher, O.; Smart, J.; Shealy, J.R.; Weimann, N.G.; Chu, K.; Murphy, M.; Schaff, W.J.; Eastman, L.F.; Dimitrov, R.; Wittmer, L.; et al. Two-dimensional electron gases induced by spontaneous and piezoelectric polarization charges in N- and Ga-face AlGaN/GaN heterostructures. *J. Appl. Phys.* **1999**, *85*, 3222–3233. [CrossRef]
31. Ibbetson, J.P.; Fini, P.T.; Ness, K.D.; DenBaars, S.P.; Speck, J.S.; Mishra, U.K. Polarization effects, surface states, and the source of electrons in AlGaN/GaN heterostructure field effect transistors. *Appl. Phys. Lett.* **2000**, *77*, 250–252. [CrossRef]
32. Higashiwaki, M.; Chowdhury, S.; Miao, M.-S.; Swenson, B.L.; Van de Walle, C.G.; Mishra, U.K. Distribution of donor states on etched surface of AlGaN/GaN heterostructures. *J. Appl. Phys.* **2010**, *108*, 063719. [CrossRef]
33. Reiner, M.; Lagger, P.; Prechtl, G.; Steinschifter, P.; Pietschnig, R.; Pogany, D.; Ostermaier, C. Modification of "native" surface donor states in AlGaN/GaN MISHEMTs by fluorination: Perspective for defect engineering. In Proceedings of the IEEE International Electron Device Meeting (IEDM), Washington, DC, USA, 7–9 December 2015.
34. Bakeroot, B.; You, S.; Wu, T.-L.; Hu, J.; Van Hove, M.; De Jaeger, B.; Geens, K.; Stoffels, S.; Decoutere, S. On the origin of the two-dimensional electron gas at AlGaN/GaN heterojunctions and its influence on recessed-gate metal-insulator-semiconductor high electron mobility transistors. *J. Appl. Phys.* **2014**, *116*, 134506. [CrossRef]
35. Matys, M.; Stoklas, R.; Blaho, M.; Adamowicz, B. Origin of positive fixed charge at insulator/AlGaN interfaces and its control by AlGaN composition. *Appl. Phys. Lett.* **2017**, *110*, 243505. [CrossRef]
36. Ťapajna, M.; Stoklas, R.; Gregušová, D.; Gucmann, F.; Hušeková, K.; Haščík, Š.; Fröhlich, K.; Tóth, L.; Pécz, B.; Brunner, F.; et al. Investigation of 'surface donors' in Al_2O_3/AlGaN/GaN metal-oxide-semiconductor heterostructures: Correlation of electrical, structural, and chemical properties. *Appl. Surf. Sci.* **2017**, *426*, 656–661. [CrossRef]
37. Ber, E.; Osman, B.; Ritter, D. Measurement of the variable surface charge concentration in Gallium Nitride and implications on device modeling and physics. *IEEE Trans. Electron Dev.* **2019**, *66*, 2100–2105. [CrossRef]

38. Esposto, M.; Krishnamoorthy, S.; Nathan, D.N.; Bajaj, S.; Hung, T.-H.; Rajan, S. Electrical properties of atomic layer deposited aluminum oxide on gallium nitride. *Appl. Phys. Lett.* **2011**, *99*, 133503. [CrossRef]
39. Ganguly, S.; Verma, J.; Li, G.; Zimmermann, T.; Xing, H.; Jena, D. Presence and origin of interface charges at atomic-layer deposited Al_2O_3/III-nitride heterojunctions. *Appl. Phys. Lett.* **2011**, *99*, 193504. [CrossRef]
40. Briere, M.A. Progress in silicon-based 600 V power GaN. *Power Semicond.* **2013**, *4*, 30.
41. Ťapajna, M.; Drobný, J.; Gucmann, F.; Hušeková, K.; Gregušová, D.; Hashizume, T.; Kuzmík, J. Impact of oxide/barrier charge on threshold voltage instabilities in AlGaN/GaN metal-oxide-semiconductor heterostructures. *Mater. Sci. Semicond. Process.* **2019**, *91*, 356–361. [CrossRef]
42. Ťapajna, M.; Kuzmík, J. Control of threshold voltage in GaN based metal–oxide–semiconductor high-electron mobility transistors towards the normally-off operation. *Jpn. J. Appl. Phys.* **2013**, *52*, 08JN08. [CrossRef]
43. Harada, N.; Hori, Y.; Azumaishi, N.; Ohi, K.; Hashizume, T. Formation of recessed-oxide gate for normally-off AlGaN/GaN high Electron mobility transistors using selective electrochemical oxidation. *Appl. Phys. Exp.* **2011**, *4*, 021002.
44. Gregušová, D.; Jurkovič, M.; Haščík, Š.; Blaho, M.; Seifertová, A.; Fedor, J.; Ťapajna, M.; Fröhlich, K.; Vogrinčič, P.; Liday, J.; et al.; et al. Adjustment of threshold voltage in AlN/AlGaN/GaN high-electron mobility transistors by plasma oxidation and Al_2O_3 atomic layer deposition overgrowth. *Appl. Phys. Lett.* **2014**, *104*, 013506. [CrossRef]
45. Hung, T.; Park, P.S.; Krishnamoorthy, S.; Nath, D.N.; Rajan, S. Interface charge engineering for enhancement-mode GaN MISHEMTs. *IEEE Electron Dev. Lett.* **2014**, *35*, 312–314. [CrossRef]
46. Blaho, M.; Gregušová, D.; Haščík, Š.; Jurkovič, M.; Ťapajna, M.; Fröhlich, K.; Dérer, J.; Carlin, J.-F.; Grandjean, N.; Kuzmík, J. Self-aligned normally-off metal-oxide-semiconductor n^{++}GaN/InAlN/GaN high-electron mobility transistors. *Phys. Status Solidi A* **2015**, *112*, 1086–1090. [CrossRef]
47. Maeda, N.; Hiroki, M.; Sasaki, S.; Harada, Y. High-temperature characteristics in recessed-gate AlGaN/GaN enhancement-mode heterostructure field effect transistors with enhanced-barrier structures. *Jpn. J. Appl. Phys.* **2013**, *52*, 08JN18. [CrossRef]
48. Lee, Y.; Kao, T.; Merola, J.J.; Shen, S. A remote-oxygen-plasma surface treatment technique for III-nitride heterojunction field-effect transistors. *IEEE Trans. Electron Dev.* **2014**, *61*, 493–497. [CrossRef]
49. Li, Z.; Chow, T.P. Channel scaling of hybrid GaN MOS-HEMTs. *Solid State Electron.* **2011**, *56*, 111–115. [CrossRef]
50. Ikeda, N.; Tamura, R.; Kokawa, T.; Kambayashi, H.; Sato, Y.; Nomura, T.; Kato, S. Over 1.7 kV normally-off GaN hybrid MOS-HFETs with a lower on-resistance on a Si substrate. In Proceedings of the 23rd International Symposium on Power Semiconductor Devices and IC's (ISPSD2011), San Diego, CA, USA, 23–26 May 2011; pp. 284–287.
51. Freedsman, J.J.; Watanabe, A.; Egawa, T. Enhancement-mode Al_2O_3/InAlN/GaN MOS-HEMT on Si with high drain current density 0.84 A/mm and threshold voltage of +1.9 V. In Proceedings of the 72nd Device Research Conference, Santa Barbara, CA, USA, 22–25 June 2014; pp. 49–50.
52. Yang, S.; Liu, S.; Liu, C.; Lu, Y.; Chen, K.J. Mechanisms of thermally induced threshold voltage instability in GaN-based heterojunction transistors. *Appl. Phys. Lett.* **2014**, *105*, 223508. [CrossRef]
53. Deen, D.; Storm, D.; Meyer, D.; Katzer, D.S.; Bass, R.; Binari, S.; Gougousi, T. AlN/GaN HEMTs with high-κ ALD HfO_2 or Ta_2O_5 gate insulation. *Phys. Status Solidi C* **2011**, *8*, 2420–2423. [CrossRef]
54. Ťapajna, M.; Kuzmík, J.; Čičo, K.; Pogany, D.; Pozzovivo, G.; Strasser, G.; Abermann, S.; Bertagnolli, E.; Carlin, J.-F.; Grandjean, N.; et al. Interface states and trapping effects in Al_2O_3- and ZrO_2/InAlN/AlN/GaN metal–oxide–semiconductor heterostructures. *Jpn. J. Appl. Phys.* **2009**, *48*, 090201. [CrossRef]
55. Čičo, K.; Hušeková, K.; Ťapajna, M.; Gregušová, D.; Stoklas, R.; Kuzmík, J.; Carlin, J.-F.; Grandjean, N.; Pogany, D.; Fröhlich, K. Electrical properties of InAlN/GaN high electron mobility transistor with Al_2O_3, ZrO_2, and $GdScO_3$ gate dielectrics. *J. Vac. Sci Technol. B* **2011**, *29*, 01A808. [CrossRef]
56. Kikkawa, T.; Makiyama, K.; Ohki, T.; Kanamura, M.; Imanishi, K.; Hara, N.; Joshin, K. High performance and high reliability AlGaN/GaN HEMTs. *Phys. Stat. Solidi A* **2009**, *206*, 1135–1144. [CrossRef]
57. Yang, S.; Huang, S.; Schnee, M.; Zhao, Q.-T.; Schubert, J.; Chen, K.J. Fabrication and Characterization of Enhancement-Mode High-κ $LaLuO_3$-AlGaN/GaN MIS-HEMTs. *IEEE Trans. Electron Dev.* **2013**, *60*, 3040–3046. [CrossRef]
58. Hsu, C.; Shih, W.; Lin, Y.; Hsu, H.; Hsu, H.; Huang, Y.; Lin, T.; Wu, C.; Wu, W.; Ma, J.; et al. Improved linearity and reliability in GaN metal–oxide–semiconductor high-electron-mobility transistors using nanolaminate La_2O_3/SiO_2 gate dielectric. *Jpn. J. Appl. Phys.* **2016**, *55*, 04EG04. [CrossRef]

59. Zhou, H.; Makiyama, K.; Ohki, T.; Kanamura, M.; Imanishi, K.; Hara, N.; Joshin, K. High-performance InAlN/GaN MOSHEMTs enabled by atomic layer epitaxy MgCaO as gate dielectric. *IEEE Electron Dev. Lett.* **2016**, *37*, 556–559. [CrossRef]
60. Huang, M.L.; Chang, Y.C.; Chang, Y.H.; Lin, T.D.; Kwo, J.; Hong, M. Energy-band parameters of atomic layer deposited Al_2O_3 and HfO_2 on $In_xGa_{1-x}As$. *Appl. Phys. Lett.* **2009**, *94*, 052106. [CrossRef]
61. Yatabe, Z.; Hori, Y.; Kim, S.; Hashizume, T. Effects of Cl_2-based inductively coupled plasma etching of AlGaN on interface properties of Al_2O_3/AlGaN/GaN heterostructures. *Appl. Phys. Express* **2013**, *6*, 016502. [CrossRef]
62. Ozaki, S.; Makiyama, K.; Ohki, T.; Okamoto, N.; Kaneki, S.; Nishiguchi, K.; Hara, N.; Hashizume, T. Effects of air annealing on DC characteristics of InAlN/GaN MOS high-electron-mobility transistors using atomic-layer-deposited Al_2O_3. *Appl. Phys. Express* **2017**, *10*, 061001. [CrossRef]
63. Ťapajna, M.; Válik, L.; Gucmann, F.; Gregušová, D.; Fröhlich, K.; Haščík, Š.; Dobročka, E.; Tóth, L.; Pécz, B.; Kuzmík, J. Low-temperature atomic layer deposition-grown Al_2O_3 gate dielectric for GaN/AlGaN/GaN MOS HEMTs: Impact of deposition conditions on interface state density. *J. Vac. Sci. Technol. B* **2017**, *35*, 01A107. [CrossRef]
64. Im, K.-S.; Ha, J.-B.; Kim, K.-W.; Lee, J.-S.; Kim, D.-S.; Hahm, S.-H.; Lee, J.-H. Normally off GaN MOSFET based on AlGaN/GaN heterostructure with extremely high 2DEG density grown on silicon substrate. *IEEE Electron Device Lett.* **2010**, *31*, 192–194.
65. Kim, K.-W.; Jung, S.-D.; Kim, D.-S.; Kang, H.-S.; Im, K.-S.; Oh, J.-J.; Ha, J.-B.; Shin, J.-K.; Lee, J.H. Effects of TMAH treatment on device performance of normally off Al_2O_3/GaN MOSFET. *IEEE Electron Device Lett.* **2011**, *32*, 1376–1378. [CrossRef]
66. Wang, Y.; Wang, M.; Xie, B.; Wen, C.P.; Wang, J.; Hao, Y.; Wu, W.; Chen, K.J.; Shen, B. High-performance normally-off Al_2O_3/GaN MOSFET using a wet etching-based gate recess technique. *IEEE Electron Device Lett.* **2013**, *34*, 1370–1372. [CrossRef]
67. Wang, M.; Wang, Y.; Zhang, C.; Xie, B.; Wen, C.P.; Wang, J.; Hao, Y.; Wu, W.; Chen, K.J.; Shen, B. 900 V/1.6 mΩ cm^2 normally off Al_2O_3/GaN MOSFET on silicon substrate. *IEEE Trans. Electron Dev.* **2014**, *61*, 2035–2040. [CrossRef]
68. Asahara, R.; Nozaki, M.; Yamada, T.; Ito, J.; Nakazawa, S.; Ishida, M.; Ueda, T.; Yoshigoe, A.; Hosoi, T.; Shimura, T.; et al. Effect of nitrogen incorporation into Al-based gate insulators in AlON/AlGaN/GaN metal–oxide–semiconductor structures. *Appl. Phys. Express* **2016**, *9*, 101002. [CrossRef]
69. Kikuta, D.; Itoh, K.; Narita, T.; Mori, T. Al_2O_3/SiO_2 nanolaminate for a gate oxide in a GaN-based MOS device. *J. Vac. Sci. Technol. A* **2017**, *35*, 01B122.
70. Lee, J.-G.; Kim, H.-S.; Seo, K.-S.; Cho, C.-H.; Cha, H.-Y. High quality PECVD SiO_2 process for recessed MOS-gate of AlGaN/GaN-on-Si metal–oxide–semiconductor heterostructure field-effect transistors. *Solid State Electron.* **2016**, *122*, 32–36. [CrossRef]
71. Khan, M.A.; Hu, X.; Sumin, G.; Lunev, A.; Yang, J.; Gaska, R.; Shur, M.S. Effects of air annealing on DC characteristics of InAlN/GaN MOS high-electron-mobility transistors using atomic-layer-deposited Al_2O_3. *IEEE Electron Device Lett.* **2000**, *21*, 63–65. [CrossRef]
72. Kambayashi, H.; Satoh, Y.; Kokawa, T.; Ikeda, N.; Nomura, T.; Kato, S. High field-effect mobility normally-off AlGaN/GaN hybrid MOS-HFET on Si substrate by selective area growth technique. *Solid State Electron.* **2011**, *56*, 163–167. [CrossRef]
73. Hua, M.; Zhang, Z.; Wei, J.; Lei, J.; Tang, G.; Fu, K.; Cai, Y.; Zhang, B.; Chen, K.J. Integration of LPCVD SiN_x Gate Dielectric with Recessed-gate E-mode GaN MIS-FETs: Toward High Performance, High Stability and Long TDDB Lifetime. In Proceedings of the International Electron Device Meeting 2016 (IEDM 2016), San Francisco, CA, USA, 3–7 December 2016; pp. 260–263.
74. Hua, M.; Wei, J.; Bao, Q.; Zhang, Z.; Zheng, Z.; Chen, K.J. Dependence of V_{TH} stability on gate-bias under reverse-bias stress in E-mode GaN MIS-FET. *IEEE Electron Dev. Lett.* **2018**, *39*, 413–416. [CrossRef]
75. Meneghesso, G.; Meneghini, M.; Bisi, D.; Rossetto, I.; Wu, T.-L.; Van Hove, M.; Marcon, D.; Stoffels, S.; Decoutere, S.; Zanoni, E. Trapping and reliability issues in GaN-based MIS HEMTs with partially recessed gate. *Microelectron. Reliab.* **2016**, *58*, 151–157. [CrossRef]
76. Hashizume, T.; Ootomo, S.; Inagaki, T.; Hasegawa, H. Surface passivation of GaN and GaN/AlGaN heterostructures by dielectric films and its application to insulated-gate heterostructure transistors. *J. Vac. Sci. Technol. B* **2003**, *21*, 1828–1838. [CrossRef]

77. Van Hove, M.; Kang, X.; Stoffels, S.; Wellekens, D.; Ronchi, N.; Venegas, R.; Geens, K.; Decoutere, S. Fabrication and performance of Au-free AlGaN/GaN-on-silicon power devices with Al_2O_3 and Si_3N_4/Al_2O_3 gate dielectrics. *IEEE Trans. Electron Dev.* **2013**, *60*, 3071–3078. [CrossRef]
78. Yang, S.; Tang, Z.; Wong, K.-Y.; Lin, Y.-S.; Lu, Y.; Huang, S.; Chen, K.J. Mapping of interface traps in high-performance Al_2O_3/AlGaN/GaN MIS-heterostructures using frequency- and temperature-dependent *C-V* techniques. In Proceedings of the 2013 IEEE International Electron Devices Meeting (IEDM), Washington, DC, USA, 9–11 December 2013.
79. Zhu, J.; Ma, X.-H.; Xie, Y.; Hou, B.; Chen, W.-W.; Zhang, J.-C.; Hao, Y. Improved interface and transport properties of AlGaN/GaN MIS-HEMTs with PEALD-grown AlN gate dielectric. *IEEE Trans. Electron Dev.* **2015**, *62*, 512–518.
80. Hashizume, T.; Kaneki, S.; Oyobiki, T.; Ando, Y.; Sasaki, S.; Nishiguchi, K. Effects of postmetallization annealing on interface properties of Al_2O_3/GaN structures. *Appl. Phys. Exp.* **2018**, *11*, 124102. [CrossRef]
81. Fritsch, J.; Sankey, O.F.; Schmidt, K.E.; Page, J.B. *Ab initio* calculation of the stoichiometry and structure of the (0001) surfaces of GaN and AlN. *Phys. Rev. B* **1998**, *57*, 15360. [CrossRef]
82. Eller, B.S.; Yang, J.; Nemanich, R.J. Electronic surface and dielectric interface states on GaN and AlGaN. *J. Vacuum Sci. Technol. A* **2013**, *31*, 050807. [CrossRef]
83. Ťapajna, M.; Kuzmík, J. A comprehensive analytical model for threshold voltage calculation in GaN based metal-oxide-semiconductor high-electron-mobility transistors. *Appl. Phys. Lett.* **2012**, *100*, 113509. [CrossRef]
84. Hasegawa, H.; Ohno, H. Unified disorder induced gap state model for insulator–semiconductor and metal–semiconductor interfaces. *J. Vac. Sci. Technol. B* **1986**, *4*, 1130–1138. [CrossRef]
85. Yatabe, Z.; Asubar, J.T.; Hashizume, T. Insulated gate and surface passivation structures for GaN-based power transistors. *J. Phys. D Appl. Phys.* **2016**, *49*, 393001. [CrossRef]
86. Jackson, C.M.; Arehart, A.R.; Cinkilic, E.; McSkimming, B.; Speck, J.S.; Ringel, S.A. Interface trap characterization of atomic layer deposition Al_2O_3/GaN metal-insulator-semiconductor capacitors using optically and thermally based deep level spectroscopies. *J. Appl. Phys.* **2013**, *113*, 204505. [CrossRef]
87. Capriotti, M.; Lagger, P.; Fleury, C.; Oposich, M.; Bethge, O.; Ostermaier, C.; Strasser, G.; Pogany, D. Modeling small-signal response of GaN-based metal-insulator- semiconductor high electron mobility transistor gate stack in spill-over regime: Effect of barrier resistance and interface states. *J. Appl. Phys.* **2015**, *117*, 024506. [CrossRef]
88. Miczek, M.; Mizue, C.; Hashizume, T.; Adamowicz, B. Effects of interface states and temperature on the *C-V* behavior of metal/insulator/AlGaN/GaN heterostructure capacitors. *J. Appl. Phys.* **2008**, *103*, 104510. [CrossRef]
89. Matys, M.; Kaneki, S.; Nishiguchi, K.; Adamowicz, B.; Hashizume, T. Disorder induced gap states as a cause of threshold voltage instabilities in Al_2O_3/AlGaN/GaN metal-oxide-semiconductor high-electron-mobility transistors. *J. Appl. Phys.* **2017**, *122*, 224504. [CrossRef]
90. Hasegawa, H.; Inagaki, T.; Ootomo, S.; Hashizume, T. Mechanisms of current collapse and gate leakage currents in AlGaN/GaN heterostructure field effect transistors. *J. Vac. Sci. Technol. B* **2003**, *21*, 1844–1855. [CrossRef]
91. Heiman, F.P.; Warfield, G. The effects of oxide traps on the MOS capacitance. *IEEE Trans. Electron Dev.* **1965**, *12*, 167–178. [CrossRef]
92. Shluger, A. Defects in oxides in electronic devices. In *Handbook of Materials Modeling*; Andreoni, W., Yip, S., Eds.; Springer Nature: Cham, Switzerland, 2019.
93. Choi, M.; Lyons, J.L.; Janotti, A.; Van de Walle, C.G. Impact of carbon and nitrogen impurities in high-κ dielectrics on metal-oxide semiconductor devices. *Appl. Phys. Lett.* **2013**, *102*, 142902. [CrossRef]
94. Choi, M.; Janotti, A.; Van de Walle, C.G. Native point defects and dangling bonds in α-Al_2O_3. *J. Appl. Phys.* **2013**, *113*, 044501. [CrossRef]
95. Sun, X.; Saadat, O.I.; Chang-Liao, K.S.; Palacios, T.; Cui, S.; Ma, T.P. Study of gate oxide traps in HfO_2/AlGaN/GaN metal-oxide-semiconductor high-electron-mobility transistors by use of ac transconductance method. *Appl. Phys. Lett.* **2013**, *102*, 103504. [CrossRef]
96. Warren, W.; Rong, F.C.; Poindexter, E.; Gerardi, G.; Kanicki, J. Structural identification of the silicon and nitrogen dangling-bond centers in amorphous silicon nitride. *J. Appl. Phys.* **1991**, *70*, 346–354. [CrossRef]

97. Zhu, J.; Hou, B.; Chen, L.; Zhu, Q.; Ling, Y.; Xiaowei, Z.; Zhang, P.; Ma, X.; Hao, Y. Threshold voltage shift and interface/border trapping mechanism in Al_2O_3/AlGaN/GaN MOS-HEMTs. In Proceedings of the 2018 IEEE International Reliability Physics Symposium (IRPS), Burlingame, CA, USA, 11–15 March 2018.
98. Henry, C.H.; Lang, D.V. Nonradiative capture and recombination by multiphonon emission in GaAs and GaP. *Phys. Rev. B* **1977**, *15*, 989–1016. [CrossRef]
99. Lax, M. Cascade capture of electrons in solids. *Phys. Rev.* **1960**, *119*, 1502–1523. [CrossRef]
100. Kirton, M.J.; Uren, M.J. Noise in solid-state microstructures: A new perspective on individual defects, interface states and low-frequency (1/*f*) noise. *Adv. Phys.* **1989**, *38*, 367–468. [CrossRef]
101. Grasser, T. Stochastic charge trapping in oxides: From random telegraph noise to bias temperature instabilities. *Microelectron. Reliab.* **2012**, *52*, 39–70. [CrossRef]
102. Taoka, N.; Kubo, T.; Yamada, T.; Egawa, T.; Shimizu, M. Experimental evidence of the existence of multiple charged states at Al_2O_3/GaN interfaces. *Semicond. Sci. Technol.* **2019**, *34*, 025009. [CrossRef]
103. Wu, T.-L.; Marcon, D.; Ronchi, N.; Bakeroot, B.; You, S.; Stoffels, S.; Van Hove, M.; Bisi, D.; Meneghini, M.; Groeseneken, G.; et al. Analysis of slow de-trapping phenomena after a positive gate bias on AlGaN/GaN MIS-HEMTs with in-situ Si_3N_4/Al_2O_3 bilayer gate dielectrics. *Solid State Electron.* **2015**, *103*, 127–130. [CrossRef]
104. Meneghini, M.; Rossetto, I.; Bisi, D.; Ruzzarin, M.; Van Hove, M.; Stoffels, S.; Wu, T.-L.; Marcon, D.; Decoutere, S.; Meneghesso, G.; et al. Negative bias-induced threshold voltage instability in GaN-on-Si power HEMTs. *IEEE Electron Dev. Lett.* **2016**, *37*, 474–477. [CrossRef]
105. Dalcanale, S.; Meneghini, M.; Tajalli, A.; Rossetto, I.; Ruzzarin, M.; Zanoni, E.; Meneghesso, G. GaN-based MIS-HEMTs: Impact of cascode-mode high temperature source current stress on NBTI shift. In Proceedings of the 2017 IEEE International Reliability Physics Symposium (IRPS), Monterey, CA, USA, 2–6 April 2017.
106. Chini, A.; Iucolano, F. Evolution of on-resistance (R_{ON}) and threshold voltage (V_{TH}) in GaN HEMTs during switch-mode operation. *Mat. Sci. Semicond. Proc.* **2018**, *78*, 127–131. [CrossRef]
107. Ťapajna, M.; Válik, L.; Gregušová, D.; Fröhlich, K.; Gucmann, F.; Hashizume, T.; Kuzmík, J. Threshold voltage instabilities in AlGaN/GaN MOS-HEMTs with ALD-grown Al_2O_3 gate dielectrics: Relation to distribution of oxide/semiconductor interface state density. In *2016 11th International Conference on Advanced Semiconductor Devices & Microsystems (ASDAM), Proceedings of the IEEE, Smolenice, Slovakia, 13–16 November 2016*; IEEE: Smolenice, Slovakia, 2016; pp. 1–4.
108. Fagerlind, M.; Allerstam, F.; Sveinbjörnsson, E.Ö.; Rorsman, N.; Kakanakova-Georgieva, A.; Lundskog, A.; Forsberg, U.; Janzén, E. Investigation of the interface between silicon nitride passivations and AlGaN/AlN/GaN heterostructures by *C*(*V*) characterization of metal-insulator-semiconductor heterostructure capacitors. *J. Appl. Phys.* **2010**, *108*, 014508. [CrossRef]
109. Huang, S.; Yang, S.; Roberts, J.; Chen, K.J. Threshold voltage instability in Al_2O_3/GaN/AlGaN/GaN metal–insulator–semiconductor high-electron mobility transistors. *Jpn. J. Appl. Phys.* **2011**, *50*, 110202. [CrossRef]
110. Ťapajna, M.; Čičo, K.; Kuzmík, J.; Pogany, D.; Pozzovivo, G.; Strasser, G.; Carlin, J.-F.; Grandjean, N.; Fröhlich, K. Thermally induced voltage shift in capacitance–voltage characteristics and its relation to oxide/semiconductor interface states in Ni/Al_2O_3/InAlN/GaN heterostructures. *Semicond. Sci. Technol.* **2009**, *24*, 035008. [CrossRef]
111. Winzer, A.; Schuster, M.; Hentschel, R.; Ocker, J.; Merkel, U.; Jahn, A.; Wachowiak, A.; Mikolajick, T. Analysis of threshold voltage instability in AlGaN/GaN MISHEMTs by forward gate voltage stress pulses. *Phys. Status Solidi A* **2016**, *213*, 1246–1251. [CrossRef]
112. Reisinger, H.; Grasser, T.; Gustin, W.; Schluandnder, C. The statistical analysis of individual defects constituting NBTI and its implications for modeling DC- and AC stress. In Proceedings of the 2010 IEEE International Reliability Physics Symposium (IRPS), Anaheim, CA, USA, 2–6 May 2010.
113. Ostermaier, C.; Lagger, P.; Prechtl, G.; Grill, A.; Grasser, T.; Pogany, D. Dynamics of carrier transport via AlGaN barrier in AlGaN/GaN MIS-HEMTs. *Appl. Phys. Lett.* **2017**, *110*, 173502. [CrossRef]
114. Lagger, P.; Steinschifter, P.; Reiner, M.; Stadtmüller, M.; Denifl, G.; Naumann, A.; Mueller, J.; Wilde, L.; Pogany, D.; Ostermaier, C. Role of the dielectric for the charging dynamics of the dielectric/barrier interface in AlGaN/GaN based metal-insulator-semiconductor structures under forward gate bias stress. *Appl. Phys. Lett.* **2014**, *105*, 033512. [CrossRef]
115. Wu, C.; Han, P.-C.; Luc, Q.H.; Hsu, C.Y.; Hsieh, T.-E.; Wang, H.-C.; Lin, Y.-K.; Chang, P.-C.; Lin, Y.-C.; Chang, E.Y. Normally-OFF GaN MIS-HEMT with F– doped gate insulator using standard ion implantation. *IEEE J. Electron Dev. Soc.* **2018**, *6*, 893–899. [CrossRef]

116. Kikkawa, T.; Hosoda, T.; Imanishi, K.; Shono, K.; Itabashi, K.; Ogino, T.; Miyazaki, Y.; Mochizuki, A.; Kiuchi, K.; Kanamura, M.; et al. 600 V JEDEC-qualified highly reliable GaN HEMTs on Si substrates. In Proceedings of the 2014 IEEE International Electron Devices Meeting (IEDM), San Francisco, CA, USA, 15–17 December 2014.
117. Kaczer, B.; Grasser, T.; Roussel, P.J.; Martin-Martinez, J.; O'Connor, R.; O'Sullivan, B.J.; Groeseneken, G. Ubiquitous relaxation in BTI stressing-New evaluation and insights. In Proceedings of the 2008 IEEE International Reliability Physics Symposium, Phoenix, AZ, USA, 24 April–1 May 2008; pp. 20–27.
118. Bisi, D.; Chan, S.H.; Liu, X.; Yeluri, R.; Keller, S.; Meneghini, M.; Meneghesso, G.; Zanoni, E.; Mishra, U.K. On trapping mechanisms at oxide-traps in Al_2O_3/GaN metal-oxide-semiconductor capacitors. *Appl. Phys. Lett.* **2016**, *108*, 112104. [CrossRef]
119. Acurio, E.; Crupi, F.; Magnone, P.; Trojman, L.; Meneghesso, G.; Iucolano, F. On recoverable behavior of PBTI in AlGaN/GaN MOS-HEMT. *Solid State Electron.* **2017**, *132*, 49–56. [CrossRef]
120. Iucolano, F.; Parisi, A.; Reina, S.; Meneghesso, G.; Chini, A. Study of threshold voltage instability in E-mode GaN MOS-HEMTs. *Phys. Stat. Solidi C* **2016**, *13*, 321–324. [CrossRef]
121. Chini, A.; Iucolano, F. Experimental and numerical analysis of V_{TH} and R_{ON} drifts in E-mode GaN HEMTs during switch-mode operation. *Mat. Sci. Semicond. Proc.* **2019**, *98*, 77–80. [CrossRef]
122. Sang, F.; Wang, M.; Zhang, C.; Tao, M.; Xie, B.; Wen, C.P.; Wang, J.; Hao, Y.; Wu, W.; Shen, B. Investigation of the threshold voltage drift in enhancement mode GaN MOSFET under negative gate bias stress. *Jpn. J. Appl. Phys.* **2015**, *54*, 044101. [CrossRef]
123. Lelis, A.J.; Habersat, D.; Green, R.; Ogunniyi, A.; Gurfinkel, M.; Suehle, J.; Goldsman, N. Time dependence of bias-stress-induced SiC Mosfet threshold-voltage instability measurements. *IEEE Trans. Electron Dev.* **2008**, *55*, 1835–1840. [CrossRef]
124. Hua, M.; Qian, Q.; Wei, J.; Zhang, Z.; Tang, G.; Chen, K.J. Bias temperature instability of normally-off GaN MIS-FET with low-pressure chemical vapor deposition SiN_x gate dielectric. *Phys. Stat. Solidi A* **2018**, *215*, 1700641. [CrossRef]
125. He, J.; Hua, M.; Zhang, Z.; Chen, K.J. Performance and V_{TH} stability in E-mode GaN fully recessed MIS-FETs and partially recessed MIS-HEMTs with LPCVD-SiN_x/PECVD-SiN_x gate dielectric stack. *IEEE Trans. Electron Dev.* **2018**, *65*, 3185–3191. [CrossRef]
126. Killat, N.; Tapajna, M.; Faqir, M.; Palacios, T.; Kuball, M. Evidence for impact ionisation in AlGaN/GaN HEMTs with InGaN back-barrier. *Electron. Lett.* **2011**, *47*, 405–406. [CrossRef]
127. Degraeve, R.; Kaczer, R.B.; Groesenenken, G. Degradation and breakdown in thin oxide layers: Mechanisms, models and reliability prediction. *Microelectron. Reliab.* **1999**, *39*, 1445–1460. [CrossRef]
128. Hua, M.; Wei, J.; Bao, Q.; He, J.; Zhang, Z.; Zheng, Z.; Lei, J.; Chen, K.J. Reverse-bias stability and reliability of hole-barrier-free E-mode LPCVD-SiNx/GaN MIS-FETs. In Proceedings of the 2017 IEEE International Electron Devices Meeting (IEDM), San Francisco, CA, USA, 2–6 December 2017.
129. Hua, M.; Cai, X.; Yang, S.; Zhang, Z.; Zheng, Z.; Wang, N.; Chen, J.K. Enhanced gate reliability in GaN MIS-FETs by converting the GaN channel into crystalline gallium oxynitride. *ACS Appl. Electron. Mat.* **2019**, *1*, 642–648. [CrossRef]
130. Hua, M.; Yang, S.; Zheng, Z.; Wei, J.; Zhang, Z.; Chen, J.K. Effects of Substrate termination on reverse-bias stress reliability of normally-off lateral GaN-on-Si MIS-FETs. In Proceedings of the 31st International Symposium on Power Semiconductor Devices & ICs, Shanghai, China, 19–23 May 2019.
131. Moens, P.; Liu, C.; Banerjee, A.; Vanmeerbeek, P.; Coppens, P.; Ziad, H.; Constant, A.; Li, Z.; De Vleeschouwer, H.; Roig-Guitart, J.; et al.; et al. An industrial process for 650V rated GaN-on-Si power devices using in-situ SiN as a gate dielectric. In Proceedings of the 2014 IEEE 26th International Symposium on Power Semiconductor Devices & IC's (ISPSD), Waikoloa, HI, USA, 15–19 June 2014; pp. 374–377.

Publisher's Note: MDPI stays neutral with regard to jurisdictional claims in published maps and institutional affiliations.

© 2020 by the author. Licensee MDPI, Basel, Switzerland. This article is an open access article distributed under the terms and conditions of the Creative Commons Attribution (CC BY) license (http://creativecommons.org/licenses/by/4.0/).

MDPI
St. Alban-Anlage 66
4052 Basel
Switzerland
Tel. +41 61 683 77 34
Fax +41 61 302 89 18
www.mdpi.com

Crystals Editorial Office
E-mail: crystals@mdpi.com
www.mdpi.com/journal/crystals

www.ingramcontent.com/pod-product-compliance
Lightning Source LLC
LaVergne TN
LVHW070555100526
838202LV00012B/470